BACTERIOLOGY RESEARCH DEVELOPMENTS

SALMONELLA

PREVALENCE, RISK FACTORS AND TREATMENT OPTIONS

Bacteriology Research Developments

Additional books in this series can be found on Nova's website
under the Series tab.

Additional e-books in this series can be found on Nova's website
under the e-book tab.

BACTERIOLOGY RESEARCH DEVELOPMENTS

SALMONELLA

PREVALENCE, RISK FACTORS AND TREATMENT OPTIONS

CHRISTOPHER B. HACKETT
EDITOR

nova publishers
New York

Library of Congress Cataloging-in-Publication Data

ISBN: 978-1-63463-651-3

Library of Congress Control Number: 2014956984

Published by Nova Science Publishers, Inc. † New York

CONTENTS

PREFACE

The ability of various microorganisms to attach to surfaces and create biofilms on them is rather a cause of concern for many industries, including for those occupied with food production and processing. Thus, the attachment of bacterial pathogens to food processing equipment is considered as an essential contributing factor in foodborne disease outbreaks, since this may ultimately lead to the contamination of food products. Improperly cleaned surfaces promote soil build-up, and, in the presence of water, contribute to the development of microbial biofilms which may contain pathogenic bacteria, such as *Salmonella. Salmonella enterica* is one of the most significant enteric foodborne bacterial pathogens, with host-adapted strains able to cause systemic human infections and persist for long periods of time, posing significant public health problems. This book discusses the prevalence, risk factors and treatment options of *Salmonella*.

Chapter 1 – Morbidity and mortality rates by *Salmonella enterica* constitute alarming concerns on a global scale. One of the many virulence factors influencing these salmonelloses is bacterial resistance to clinically used chemotherapeutic antimicrobial agents. Thus, clinical isolates of drug and multidrug resistant *S. enterica* represent a particularly serious public health concern. Mechanisms of bacterial drug resistance include (a) biofilm formation (b) enzymatic destruction of antibacterial drugs (c) reduced intracellular drug permeability (d) drug target alteration and (e) active drug efflux. Among these antibacterial drug resistance mechanisms, active efflux of chemotherapeutic drugs from *S. enterica* represents a key resistance mechanism. The classes of active antimicrobial efflux transporters harbored by *S. enterica* include (i) the ABC (ATP binding cassette) family of transporters (ii) the MATE (multi antimicrobial extrusion or multidrug and toxic compound extrusion) superfamily of transporters (iii) the RND (resistance nodulation division) superfamily of transporters (iv) the SMR (small multidrug resistance) superfamily of transporters and (v) the MFS (major facilitator superfamily) of transporters. Among these active drug efflux systems, the MFS represents one of the largest superfamilies of transporters and is a major class of multidrug resistance in *S. enterica*. This review chapter explores the molecular biology of the drug and multidrug efflux pumps harbored by *S. enterica*. Future directions include modulation of the antimicrobial efflux pumps in order to circumvent clinical multidrug resistance in *S. enterica* and eventually restore clinical drug efficacy.

Chapter 2 – *Salmonella enterica* serovar Heidelberg is one of the most commonly detected serovars in swine and poultry, ranking among the top five serotypes associated with human salmonellosis. Although *S.* Heidelberg outbreaks have been linked with several foods of animal origin, poultry-derived foods serve as the major source of human infections in the United States and Canada. Moreover, recent research has shown that this serotype of *Salmonella* is significantly more invasive in humans when compared to other non-typhoidal counterparts. Severe infections caused by *S.* Heidelberg require antibiotic therapy in both adults and children. However, *S.* Heidelberg strains resistant to multiple antibiotics, including streptomycin, kanamycin, ceftriaxone, ciprofloxacin and other beta-lactam antibiotics, have been isolated from foods, especially poultry meat, thereby raising concerns on poultry products serving as a reservoir of multidrug resistant food-borne salmonellosis in humans. This chapter discusses the emergence of *S.* Heidelberg as a significant foodborne pathogen with emphasis on its prevalence, virulence, antibiotic resistance, and public health significance. Additionally, potential strategies for controlling *S.* Heidelberg in poultry are highlighted.

Chapter 3 – *Salmonella* is a foodborne pathogen that causes a huge amount of cases of typhoid fever, gastroenteritis, and deaths every year throughout the world. Although salmonellosis cases in humans have decreased in the last five years, *Salmonella* remains the second most common zoonosis in humans. Foodborne outbreaks caused by *Salmonella* have also reduced in recent years, but they have been linked with contamination of eggs and egg products, cheese, mixed foods, and fresh fruits and vegetables. Therefore control measures for this microorganism are very important to prevent and control *Salmonella* at relevant stages of production, processing, and distribution, especially in primary production, thus reducing its prevalence and the risk it poses to public health. In this context, research carried out to find antimicrobial compounds from natural sources is important because they could be used as additives in new product formulations, where they could exercise an additional measure to control *Salmonella* growth and have an important impact from economic and food safety points of view.

By-products from the food industry are a potential source of inexpensive raw materials, and are rich in bioactive components whose technological and antimicrobial properties are still scarcely studied. With the aim of covering this gap, the objective of the present study was focused on evaluating the antimicrobial properties of three citrus by-products – mandarin, orange, and lemon – against *Salmonella enterica* serovar Typhimurium, in reference medium, under various incubation conditions with differences in temperature and by-product concentration. According to the results obtained, it can be concluded that all the citrus by-products showed a bacteriostatic and/or bactericidal effect under the conditions studied, the mandarin by-product being the most effective one. Maximum reduction levels in the microbial population attained values of $\approx 8\ \log_{10}$ cycles at refrigeration temperature (5 °C). Consequently, it can be concluded that citrus by-products have effective antimicrobial activity, and could act as an additional barrier to microbial growth when added to pasteurized beverages that are stored under refrigeration, contributing additionally to meeting the zero waste targets set by the European Union.

Chapter 4 – A specific strain of *S.* Enteritidis (SE86) was responsible for more than 95 % of the investigated salmonellosis occurred in the State of Rio Grande do Sul (RS), Southern Brazil, from 1999 to 2013. The aim of this chapter is to demonstrate the involvement of this strain with the salmonellosis outbreaks and to discuss the factors that probably contributed to *S.* Enteritidis SE86 had become one of the most important food pathogens in Southern Brazil. During 1999 to 2006, 190 salmonellosis outbreaks were investigated in RS and one DNA banding profile was identified among *S.* Enteritidis isolated from foods, blood and feces of victims. The causative strain was named *S.* Enteritidis SE86. The main risk factors for the outbreaks caused by SE86 were 1) the consumption of homemade mayonnaise prepared with raw eggs; 2) holding foods in room temperature for more than 2 hours, 3) cross-contamination due to the contact of food with contaminated equipment and utensils. The growth of SE86 in homemade mayonnaise was modeled and results demonstrated that SE86 was able to grow faster than others *Salmonella* serovars, during the first six hours at environmental temperature; however SE86 did not grow in homemade mayonnaise at 10° C, during 18 hours. Further studies have demonstrated that SE86 was able to form biofilms on stainless steel, stainless steel welds and polypropylene surfaces, and survived to 400 and 200 ppm sodium hypochlorite. After sodium hypochlorite exposure, SE86 expressed RpoS and Dps proteins, which are involved with oxidative stress. Due to exposure to sub-lethal pH, SE86 became acid-adapted and increased its thermal resistance. Acid-adapted SE86 was able to survive to simulated gastric fluid pH 1.5 and became more virulent than other *Salmonella* serovars for germ-free mice. Isolates of SE86 remain sensitive to several antibiotics, nevertheless resistance to ampicillin and nalidixic acid and multidrug resistance has increased during last decade.

Chapter 5 – Human food- borne salmonellosis, an illness caused by *Salmonella* species, is an infection of serious public health concern as it is increasing worldwide and appears to be the primary cause of confirmed foodborne outbreaks. It is estimated that tens of millions of cases occur worldwide every year and the disease results in more than hundred thousand deaths. *Salmonella* in food is mainly detected in meat and its products but the numbers of outbreaks caused by vegetables and products thereof have increased compared to previous years. Reported outbreaks caused by consumption of fresh products such as tomatoes, cucumbers, watermelons, cantaloupes, sprouts and juices have been linked to *Salmonella.* Meanwhile *Salmonella* has been isolated from a variety of fresh products and its prevalence has been reported to range from 0.6% to 72% in strawberries, fennel, lettuce, oranges, cantaloupes, onions, sprouts etc. In this chapter prevalence of *Salmonella* in unprocessed and minimally processed fruits and vegetables as well as salmonellosis outbreaks linked to the consumption of such products are addressed.

Chapter 6 – The ability of various microorganisms to attach to surfaces and create biofilms on them is rather a cause of concern for many industries, including for those occupied with food production and processing. Thus, the attachment of bacterial pathogens to food processing equipment is considered as an essential contributing factor in foodborne disease outbreaks, since this may ultimately lead to the contamination of food products. Improperly cleaned surfaces promote soil build-up, and, in the presence of water, contribute to the development of microbial biofilms which may contain pathogenic bacteria, such as *Salmonella*. It is well recognized that biofilm cells differ physiologically from their planktonic counterparts, presenting a modified and heterogeneous gene expression profile. Additionally, it has been observed that the resistance of sessile cells to antimicrobials and other environmental stresses is significantly increased compared to what is normally seen with the same cells being planktonic. Noteworthily, salmonellae have been shown to survive for years in non-enteric habitats, including sessile communities on food contact and product surfaces. Indeed, several reports have demonstrated the ability of *Salmonella* to attach and form biofilms on abiotic surfaces, such as stainless steel, plastic, rubber, glass, marble and cement. *Salmonella* is also able to strongly attach and persist on both animal and plant (produce) surfaces. It is believed that the attachment to all these surfaces and the subsequent biofilm formation on them enhance the capacity of pathogenic *Salmonella* bacteria to successfully cope with hurdles that are commonly encountered outside the host and within food processing. The purpose of this chapter is to review the current available knowledge related to the attachment of *Salmonella* to food contact and product surfaces and the possible subsequent sessile development on them in view of the strong impact of these two interrelated capabilities on the enhancement of its survival outside the host, its environmental persistence and spread. Undoubtedly, the ability to recognize why and how *Salmonella* attach to such surfaces is an important area of focus, since this may provide valuable ways towards the elimination of this important pathogen from food processing environments and eventually lead to reduced *Salmonella*-associated human illness.

Chapter 7 – *Salmonella* colonize the intestinal tract of a wide variety of animals. Within the intestinal tract, *Salmonella* tend to be scavengers or secondary degraders that are dependent upon nutrients released by the enzymatic action of the more abundant primary degraders which digest intestinal glycoproteins and complex carbohydrates. *Salmonella* are catalytically robust and capable of using a wide variety of carbon and nitrogen sources. While a considerable amount of information is available on *Salmonella* pathogenicity genes and how they facilitate infection, the authors have only recently begun to appreciate the nutritional requirements of *Salmonella* within the context of the intestinal microbiome. Several recent studies have used –omics driven (i.e., genomics, transcriptomics, proteomics, metabolomics) approaches to investigate possible contributions of metabolic processes to *Salmonella* colonization and virulence. The authors review here recent research literature on the roles of metabolic genes in *Salmonella* colonization in various host systems including mice, swine, calves and poultry, as well as metabolic genes implicated in the persistence of *Salmonella* on produce. Understanding the nutritional requirements of *Salmonella* in the host intestinal tract is important for identifying potential targets and designing new approaches for reducing growth of *Salmonella* during the food animal production period and thereby mitigating the entry of *Salmonella* into the human food supply.

Chapter 8 – Salmonella is the confirmed causative agent of numerous foodborne outbreaks throughout the world. Successful monitoring of the outbreaks and effective implementation of control programs essentially requires the accurate identification of the infection source and the transmission pathways throughout the food chain. Traditional typing relied on phenotypic characteristics such as serotyping, phage typing and antimicrobial susceptibility testing. More recently a wide range of methods relying on differences at genome level, such as Pulsed-Field Gel Electrophoresis (PFGE), Multi-Locus Variable number of tandem repeats Analysis (MLVA), Multi Locus Sequence Typing (MLST), Single Nucleotide Polymorphisms (SNPs) of Whole Genome Sequences (WGS) and Clustered Regularly Interspaced Palindromic Repeats (CRISPR) have been introduced, thoroughly studied and extensively applied. All these techniques are characterized by certain advantages and drawbacks that should be taken into consideration before any conclusion of epidemiological nature is drawn. In this chapter all information relevant to current typing schemes of food-associated salmonellae is integrated and critically reviewed.

Chapter 9 – *Salmonella* is a gram-negative microorganism, generally dispersed in nature and habitually found in the intestinal tract of animals and humans. *Salmonella* reveals a considerable aptitude to adapt to physical and chemical stresses in the environment. Survival mechanisms are stimulated following the detection of environmental signals and provoke a complicated adaptive response that guides to a state of tolerance and thus survival below sub-optimal or even sub-lethal conditions. This microorganism can persist for long periods in the environment in a heavily stressed state known variously, and often contentiously, as viable but nonculturable.

Salmonella as many other bacteria resist to environmental stress by adopting the viable but nonculturable phenotype (VBNC). VBNC bacteria have been exemplified as organisms that fail to grow and develop colonies on routine bacteriological media, but which accrue to be considered alive since of their aptitude of metabolic activity and preserve virulence.

Is it convincing that the VBNC phenotype can revert to a culturable state, and vice versa.

This chapter presents and evaluates the hypotheses concerning the resistance of VBNC forms of *Salmonella* under extended period of time of stressful conditions.

In: Salmonella
Editor: Christopher B. Hackett

ISBN: 978-1-63463-651-3

Chapter 1

MULTIDRUG RESISTANCE EFFLUX PUMPS OF *SALMONELLA ENTERICA*

Ugina Shrestha[1], Indrika Ranaweera[1], Sanath Kumar[2], Ranjana K. C.[1], Prathusha Kakarla[1], Wazir Singh Lakra[2], Guixin He[3], Jody Andersen[1] and Manuel F. Varela[1*]

[1]Eastern New Mexico University, Department of Biology, Portales, New Mexico, US

[2]QC Laboratory, Harvest and Post Harvest Technology Division, Central Institute of Fisheries Education (CIFE), Seven Bungalows, Versova, Andheri (W), Mumbai, India

[3]University of Massachusetts Lowell, Department of Clinical Laboratory and Nutritional Sciences, Lowell, MA, US

ABSTRACT

Morbidity and mortality rates by *Salmonella enterica* constitute alarming concerns on a global scale. One of the many virulence factors influencing these salmonelloses is bacterial resistance to clinically used chemotherapeutic antimicrobial agents. Thus, clinical isolates of drug and multidrug resistant *S. enterica* represent a particularly serious public health concern. Mechanisms of bacterial drug resistance include (a) biofilm formation (b) enzymatic destruction of antibacterial drugs (c) reduced intracellular drug permeability (d) drug target alteration and (e) active drug efflux. Among these antibacterial drug resistance mechanisms, active efflux of chemotherapeutic drugs from *S. enterica* represents a key resistance mechanism. The classes of active antimicrobial efflux transporters harbored by *S. enterica* include (i) the ABC (ATP binding cassette) family of transporters (ii) the MATE (multi antimicrobial extrusion or multidrug and toxic compound extrusion) superfamily of transporters (iii) the RND (resistance nodulation division) superfamily of transporters (iv) the SMR (small multidrug resistance) superfamily of transporters and (v) the MFS (major facilitator superfamily) of transporters. Among these active drug efflux systems, the MFS represents one of the largest superfamilies of transporters and is a major class of multidrug resistance in *S.*

* Corresponding author. Email: manuel.varela@enmu.edu.

enterica. This review chapter explores the molecular biology of the drug and multidrug efflux pumps harbored by *S. enterica*. Future directions include modulation of the antimicrobial efflux pumps in order to circumvent clinical multidrug resistance in *S. enterica* and eventually restore clinical drug efficacy.

Keywords: *Salmonella enterica*, active drug efflux, major facilitator superfamily, multidrug resistance, antibiotic resistance, salmonellosis

Medical Importance of Salmonellosis

Salmonella enterica is a Gram-negative facultatively-anaerobic, rod-shaped bacterium that is motile, non-endospore forming, and 0.7 - 1.5 by 2.0 - 5.0 μm in size. *S. enterica* is a member of the Enterobacteriaceae family of Gram-negative rods [1, 2]. *Salmonella* serotypes differ in their extracellular antigens, natural reservoirs, and their ability to cause human infections [3-5]. *S. enterica* may live in the gastrointestinal tracts of humans and non-human animals as a part of the normal microbiome without causing clinical disease, but may be excreted, leading to transmission and infection of other individuals [6]. *S. enterica* may be found in foods as a result of fecal contamination by a carrier or an infected individual [7], and are usually transmitted to others by ingestion of food or water contaminated by infected feces.

Food borne *Salmonella* infections, i.e., salmonelloses, are a major clinical problem worldwide [8]. *S. enterica* is estimated to be a leading cause of hospitalizations and mortality from foodborne diseases [9]. The four main clinical manifestations of *Salmonella* infections are gastroenteritis, bacteremia, enteric fever and asymptomatic colonization / carrier state [10, 11]. Children are the most likely to get infected with *Salmonella* and develop salmonellosis. Children under five years of age exhibit the highest rates of salmonellosis. At the same time, the rate of infection has seasonal changes as it is more common in the summer than in winter. Salmonellosis is more common in children under the age of 5, adults 20-30 year olds, and patients 70 years or older [10]. Only about 50 serotypes among the recognized serovars of *S. enterica* are isolated in any significant numbers as human pathogens [12, 13].

Of these *Salmonella* pathogens, most cause acute gastroenteritis, which is the most common form of salmonellosis. Also known as food poisoning, gastroenteritis is characterized by a short incubation period and a predominance of an initial presentation of symptoms such as nausea, vomiting, diarrhea, fever, headache, abdominal cramps, and myalgias [10, 11, 13-16]. The symptoms can range from mild to severe and may persist for 2-7 days [10, 11]. Typically, *S. enterica* serovar Typhimurium is the most common serotype which can cause a severe form of gastroenteritis in humans, characterized by colonic involvement, septicemia, fever, and or abortion [11, 13, 15, 16].

A second common clinical manifestation is typhoid fever, also known as enteric fever, and it is caused by *S.* Typhi, which is the causative agent of typhoid fever. This serotype of *Salmonella* is a major health problem in developing countries. Transmission of the disease is via the fecal-oral route [17], and a milder form of this disease is caused by serotypes Paratyphi A, B, and C [10, 18]. This febrile illness is characterized by gradually increasing fever, rising within 72 hours of the onset of illness with complaints of headache, abdominal pain, anorexia, malaise, myalgias. These symptoms can persist for about 1 week or longer and are then followed by gastrointestinal symptoms of diarrhea and vomiting [11]. The symptoms can be accompanied by a faint rose-colored rash on the abdomen and chest with less common symptoms such as sore throat, cough, and bloody diarrhea [10, 19]. Enteric fever, however, is a serious clinical disease characterized by a gradual onset of symptoms, interstitial inflammation, and, ultimately, bacterial dissemination to systemic sites [20]. Complications can occur like myocarditis, intestinal segmental ileus, encephalopathy, intravascular coagulation, infection of the biliary tree and intestinal tract, urinary tract infection, plus metastatic lesions in bone, joints, liver, and meninges, and neuropsychiatric manifestations [19]. The most severe complications is hemorrhage due to perforations of the terminal ileum of the proximal colon, which occur in 3% of patients. Fever can last for weeks if untreated. However with proper antimicrobial therapy, seriously ill patients usually convalesce within 10 - 14 days [2, 10, 19].

Salmonella serovars Typhi and Paratyphi (A, B and C) can penetrate the intestinal wall and enter the body fluids, with potential of causing severe systemic infections. The other serovars (non-Typhi or Paratyphi-NTP *Salmonella*) mostly cause food poisoning and enteritis and are commonly isolated from stool samples [21] . *S.* Typhi and *S.* Typhimurium use similar mechanisms for host cellular invasion and intracellular trafficking in cultured human epithelial cells [22].

The causative agents of bacteremia, another serious clinical manifestation, are non-typhoidal *Salmonella* serotypes Typhimurium and Enteritidis, which cause food poisoning and gastroenteritis and which are mostly acute in nature, self-limited and do not require antimicrobial therapy [21]. Nonetheless, approximately 5% of patients with gastrointestinal illness caused by non-typhoidal *Salmonella* serotypes lead to the development of bacteremia, a serious complication [23]. Among all, *Salmonella* serotype Choleraesuis mostly tend to cause bacteremia in humans. The risk for *Salmonella* bacteremia is higher in children and adults with certain underlying conditions, such as very young age, immunocompromised patients with HIV infections, AIDS, sickle cell disease, malignancies, immunosuppressive therapy, hemolytic anemia, congenital immunodeficiencies, and inflammatory bowel disease [24-27]. Bacteremia cases in these higher risk groups are more likely to develop extraintestinal focal infections such as meningitis, septic arthritis, osteomyelitis, cholangitis and pneumonia [23-26, 28-30] . In adults, the complications can be even more severe such as in the development of infectious endarteritis [31, 32]. Bacteremia occurring in immunosuppressed individuals and patients with comorbid medical conditions can cause septic shock, septicemia and eventually death [10, 19]. Thus, in these individuals, antimicrobial therapy would be indicated.

Asymptomatic colonization involves little or no symptoms. The *Salmonella* infection can be colonized in the gastrointestinal tract with associated fecal shedding, which becomes the source of infection for other animals and humans [33]. *Salmonella* strains responsible for typhoid and paratyphoid fevers often have their origins in prior human colonization [11]. Several factors such as acid intolerance, resistance to antimicrobial peptides and bile salts, and uptake and replication of *Salmonella* in the intestinal mucosa play important roles in *Salmonella* colonization in mammals [34, 35]. Chronic colonization can develop in 1% to 5% of patients; and in these patients, the gall bladder acts as a reservoir for *Salmonella* colonization [36].

EPIDEMIOLOGY

The incidences of *Salmonella* infections in humans have dramatically increased throughout the world over the past few years leading to its emergence as an important public health concern. *S. enterica has* become the second most common bacterial cause of foodborne gastroenteritis cases worldwide [37]. According to the Centers for Disease Control and Prevention (CDC), approximately 42,000 salmonellosis cases are reported every year in the United States [38]. The actual number, however, may be greater than that reported because of undiagnosed or unreported milder cases. Thus, every year, *Salmonella* infection is estimated to cause about 1.4 million human cases in the United States and to be the leading cause of 23,000 hospitalizations and 450 deaths from foodborne disease [39, 40]. The global health impact is as high as an estimated 93.8 million illnesses and 155,000 deaths each year [41].

According to the current view of taxonomy, members of genus *Salmonella* are assigned to two species: *S. enterica* and *S. bongori* where *S. enterica* itself is divided into six subspecies, denoted as *enterica*, *salamae*, *arizonae*, *diarizonae*, *indica*, and *houtenae*, or as I, II, IIIa, IIIb, IV, and VI, respectively [42]. Although over 2300 serovars of *Salmonella* have been identified, only about 50 are considered pathogens [43]. Of these, only a relatively small number of serotypes such as *S.* Typhimurium (19%), *S.* Enteritidis (14%), *S.* Newport (9%) and *S.* Javiana (5%), account for about half of all human clinical isolates reported in the U.S. Therefore, *S.* Typhimurium and *S.* Enteritidis are the most common in the U.S. [44].

Salmonella microorganisms live in the animal gastrointestinal tract, which is a common source of transmission [45]. The most commonly isolated serotype, *S.* Enteritidis, is often associated with eggs and poultry [46] and are often responsible for salmonellosis outbreaks [47, 48]. Consumption of contaminated vegetables, fruits, milk, meat and meat products is a major cause in the emergence of food-borne salmonellosis [49]. Since 2012, one of five high-priority goals for the U.S. Department of Health and Human Services has become the reduction of *S.* Enteritidis infections. Towards this, it is important to study the mechanisms of virulence and antimicrobial resistance in order to find the means to circumvent these mechanisms, to reduce the conditions that foster dissemination through populations and to restore efficacy of chemotherapy.

VIRULENCE MECHANISMS OF *S. ENTERICA*

The virulence factors associated with *S. enterica* have been extensively studied. Salmonellosis begins with exposure and consumption of *S. enterica*; initially, the acid tolerance response (ATR) is invoked, allowing survival of the bacterium in the low pH of the stomach. The ATR mechanism maintains the internal pH homeostasis to ensure bacterial viability [50]. The pH stress response in *S. enterica* is complex. Thus far, about 51 acid shock proteins (ASPs) associated with *S. enterica* have been identified [51]. Upon reaching the gastrointestinal tract the bacteria enter the M cells, thus circumventing the intestinal epithelial barrier [52]. Adhesion of the bacterium to cells of the Peyer's patches is mediated by gene products of the fimbriae *lpfABCDE* operon [53, 54]. Subsequent to bacterial adhesion, the host cell immediately undergoes cytoskeletal changes which are required for bacterial entry [55]. The host cell surface membrane ruffles during pinocytosis, forming *Salmonella* containing vacuoles (SCV) in which the enclosed bacteria multiply [56]. *S. enterica* enclosed within M cells may avoid fusion with internal hydrolytic lysosomes and become released, eventually encountering macrophages residing in nearby tissues after crossing the lamina propria of the intestinal wall [57]. Subsequent bacterial multiplication inside the macrophage is facilitated by the virulence factor SopD [58], allowing dissemination to other tissues [59]. Additionally, a variety of virulence factors are delivered to the host cell via a Type III Secretion System (T3SS) harbored by *S. enterica* [60, 61].

Chemotherapy of Clinical Salmonelloses

Antimicrobial chemotherapy for the treatment of salmonellosis is less common since a study by Wiström and Norrby in 1995 found little to no benefit from the administration of fluoroquinolones, which were beneficial only when norfloxacin was administered within 48 hours after the onset of symptoms [62]. It was once thought that antimicrobial agents would shorten the duration of *S. enterica* infections; presently, however, electrolyte and fluid replacement is the preferred treatment avenue as the majority of relatively milder infections are self-limiting. Antibiotics are indicated, however, in patients who are severely ill, and the most commonly preferred drugs are the fluoroquinolones, trimethoprim sulfamethoxazole (TMP-SMZ), ampicillin, or expanded- spectrum cephalosporins, such as ceftriaxone or cefixime [37, 63, 64] . Regretfully, multidrug resistance has already been documented in a large number of *S.* Typhimurium isolates, particularly to TMP-SMZ and ampicillin, thus confounding the chemotherapeutic efficacy in the treatment of severely ill patients [65-68]. Clinical isolates from a 2008 salmonellosis outbreak involving 43 states of the U.S. [69] were largely sensitive to antimicrobial agents but with a multidrug resistant *S.* Typhimurium DT104 isolate being an exception [70].

Antibiotic Resistance Conferring Mechanisms on the *Salmonella enterica* Genome

Many cellular and molecular systems in bacteria work together to impart antibiotic resistance upon pathogenic strains [71-73]. Extensive use of antimicrobial agents has led to the dissemination of multidrug resistance (MDR), leading to the notion that resistance mechanisms represent virulence factors [47, 74, 75]. The drug resistance mechanisms like (a) biofilm formation, (b) ribosomal protection, (c) modification of drug binding sites, (d) enzymatic-inactivation of the antibiotics, (e) decreased membrane permeability and (f) active drug efflux pumps all contribute to resistance against multiple structurally diverse antimicrobial agents in bacteria [72, 76-78] (Figure 1). Various resistance mechanisms encoded on the genome of Salmonella enterica function by extruding antibiotics out of the bacterial cell to dilute the cytoplasmic drug levels and confer MDR [47].

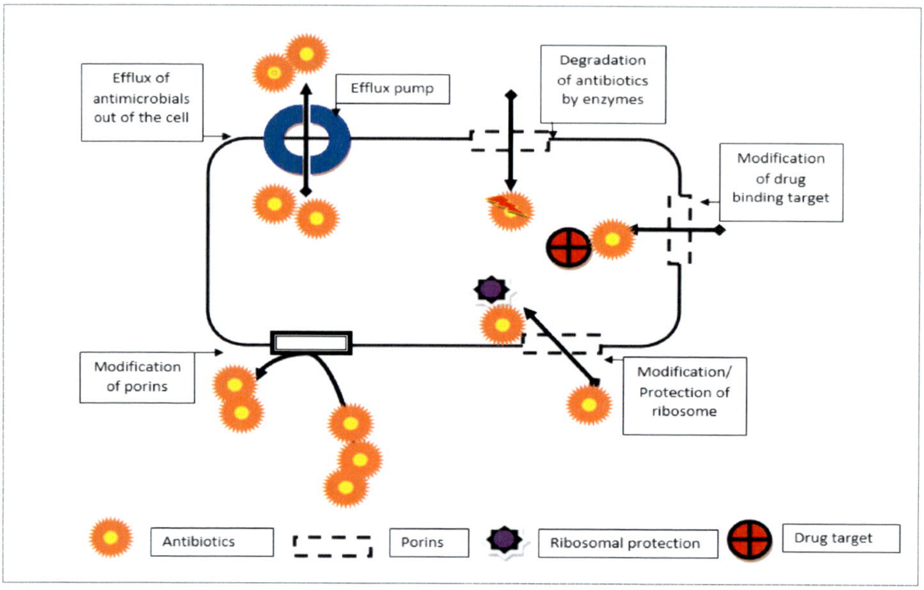

Figure 1. Bacterial antimicrobial resistance mechanisms. The various molecular mechanisms of bacterial resistance to antimicrobial drugs include: efflux of antibiotics; degradation of antibiotics by *Salmonella*-secreted enzymes; modification of drug targets such as the ribosome; and modification or removal of porins, decreasing the membrane permeability.

Drug and Multidrug Efflux Pumps of *Salmonella enterica*

Drug and multidrug efflux pumps represent a major mechanism of bacterial resistance to antimicrobial agents [79]. Antimicrobial efflux pumps from *Salmonella* actively extrude a variety of antimicrobials across the biological membrane and out of the cell, protecting the bacterium from the growth inhibitory effects of anti-bacterial substances [47, 49]. Empirical data suggest the role of active efflux pumps in Salmonella as conferring resistance to various antimicrobial agents [80]. The classes of active drug efflux transporters harbored by S. enterica include: (i) the ABC (ATP binding cassette) transporters [81], (ii) the MATE (multi antimicrobial extrusion or multidrug and toxic compound extrusion) superfamily [73, 82], (iii) the RND (resistance nodulation division) superfamily [83], (iv) the SMR (small multidrug resistance) superfamily [84], and **(v)** the MFS (major facilitator superfamily) [85]. About twelve efflux pumps from four of the major families have been identified in *Salmonella* (Table 1) [47, 86]. In contrast, relatively little is known regarding the involvement of the SMR transporters from *Salmonella* [87].

The RND Family of Efflux Pumps

Efflux pumps from the resistance nodulation division (RND) family play a major role in producing both the intrinsic and the elevated levels of resistance to a very large range of toxic compounds in *Salmonella* spp. [88]. The tripartite drug efflux system functions together with communication between an RND efflux pump, a membrane fusion protein (MFP), and an outer membrane factor (OMF) [89].

Table 1. Efflux pumps from major transporter families, their substrates and genetic location in *Salmonella*

Gene	Pump	Family	Substrates	Location
acrA/ acrB/ tolC	AcrAB	RND	Ciprofloxacin, chloramphenicol, tetracycline, acriflavine, fusidic acid, novobiocin, erythromycin-A, rifampin, b -Lactam antibiotics, ethidium bromide, bile salts, crystal violet, sodium dodecyl sulfate (SDS), sodium deoxycholate, sodium cholate, triton X-100, cetrimide, norfloxacin, cetyl-trimethylammonium bromide (CTAB), nalidixic acid, penicillin-G and triclosan	Chromosome
acrD	AcrD	RND	Kanamycin, tobramycin, gentamicin, aminoglycosides, taurocholate, novobiocin, SDS, DOC and amikacin	Chromosome
acrEF	AcrEF	RND	Nalidixic acid, ciprofloxacin, tetracycline, chloramphenicol, triclosan and C-TAB	Chromosome
mdsA/ mdsB/ tolC/ mdsC	MdsABC	RND	Novobiocin, acriflavine, crystal violet, ethidium bromide, methylene blue, rhodamine 6G, tetraphenylphosphonium bromide (TPP), benzalkonium bromide and SDS	Chromosome
mdtA/ mdtB/ mdtC/ tolC	MdtABC	RND	β-lactams, novobiocin, SDS, DOC, metal resistance (Zinc, copper, tungstate) and bile salts	Chromosome
macA/ macB/ tolC	MacAB	ABC	Macrolides, erythromycin and detoxification of reactive oxygen species (ROS)	Chromosome
emrA/ emrB/ tolC	EmrAB	MFS	Uncouplers, nalidixic acid, novobiocin and DOC	Chromosome
mdfA	MdfA	MFS	Doxorubicin, TPP, Ethidium, Chloramphenicol, Thiamphenicol, Tetracycline, Puromycin, Pyronin, Ciprofloxacin, Norfloxacin, Benzalkonium, Erythromycin, Neomycin, IPTG and alkali tolerance	Chromosome
mdtK	MdtK	MATE	Norfloxacin, doxorubicin and acriflavine	Chromosome
tetA(A) tetA(B) tetA(G)	TetA TetB TetG	MFS	Tetracyclines	SGI-1
floR	FloR	MFS	Florfenicol	SGI-1
cmlA	CmlA	MFS	Chloramphenicol	SGI-1

THE ACRAB-TOLC EFFLUX SYSTEM

Different genotypes of *S.* Typhimurium, namely *tolC::kan*, *acrB*::*kan* and *acrAB* variants, stably express the AcrAB-TolC efflux system in their biological membranes [80]. The AcrAB efflux pump is encoded by the *acrAB* genes located on the chromosome of *S.* Typhimurium (Table 1) [88]. The AcrAB-TolC antimicrobial efflux system functions as a tripartite complex (AcrA/AcrB/TolC); AcrB is an RND efflux pump. AcrA is a periplasmic membrane fusion protein, and TolC is the outer-membrane channel. The acriflavine resistance protein B (AcrB) efflux pump contains twelve transmembrane (TM) α-helices constituting the transmembrane domains [89]. Domains TM4 and TM10 play an essential role in proton transport [89]. The periplasmic part of AcrB consists of the funnel-like TolC docking domain with a central pore formed by three α-helices. AcrB also contains a cleft at the periplasmic domain to accommodate AcrA [89]. The AcrB pump is a drug/proton antiporter with high affinities for conjugated bile salts, and AcrA strongly stimulates the activity of the AcrB pump [90]. Protonation induces a conformational change in the periplasmic domain of the AcrAB pump thereby extruding the drug using the proton motive force [89, 90]. The tripartite AcrAB-TolC efflux system of *S.* Typhimurium actively extrudes ciprofloxacin, chloramphenicol, tetracycline, acriflavine, fusidic acid, novobiocin, erythromycin, rifampin, β-lactam antibiotics, ethidium bromide, bile salts, crystal violet, sodium dodecyl sulfate (SDS), sodium deoxycholate, sodium cholate, triton X-100, cetrimide, norfloxacin, cetyl trimethylammonium bromide (CTAB), nalidixic acid, penicillin-G, and triclosan (Table 1) [88, 91]. The lipophilicity of the antibiotic side chains constitute a basic structural requirement in order for penicillins and cephalosporin to serve as substrates for the AcrAB pump system [88]. Over-expression of the AcrAB-TolC efflux system leads to a high-level of resistance to nalidixic acid, fluoroquinolones, such as enrofloxacin, ciprofloxacin, and marbofloxacin, and to structurally unrelated drugs like flumequine, vetoquinol, chloramphenicol, florfenicol, and tetracycline in isolates of *S.* Typhimurium DT204 from Belgium and Germany [92]. Over-expression from the *acrA*, *acrB* and *tolC* genes lead to fluoroquinolone resistance and efflux of substrates like chloramphenicol, nalidixic acid, ciprofloxacin and tetracycline in an MDR strain of *S.* Typhimurium [49, 80]. Active efflux by AcrA is an additional high-quinolone resistance conferring mechanism that increases resistance to florfenicol even in the absence of the specific *floR* resistance gene [92]. Elevated expression of AcrA is associated with quinolone and fluoroquinolone resistances in *Salmonella* mutants and with induction of MDR in *Salmonella* cells that are tolerant to organic solvents [91, 92]. Inactivation of the *acrB* gene coding for the multidrug transporter of the AcrAB-TolC efflux system resulted in a hyper-susceptible phenotype to quinolones, enrofloxacin, chloramphenicol, florfenicol, and tetracycline [92]. Mutations in gyrase or topoisomerase IV along with AcrB over-expression was shown to induce high clinical fluoroquinolone tolerance leading to failure in fluoroquinolone therapy [91]. The AcrAB-TolC pump is transcriptionally regulated by the *marRAB* and *soxRS* systems of *S. enterica* [91]. It was shown that indole and bile salts induce AcrAB expression in *Salmonella* via RamA and AraC [80]. Phenyl-arginine beta-naphthylamide (PAβN) was identified as an effective efflux pump inhibitor (EPI) of the AcrAB-TolC efflux system in *S.* Typhimurium. This inhibitor mimics and competitively blocks the substrate-binding pocket within the AcrB efflux pump [80].

THE ACRD EFFLUX PUMP

AcrD and AcrEF (see below) are the other RND efflux pumps encoded on the *S. enterica* chromosome [91]. AcrD is homologous to AcrAB and AcrEF. It was shown that deletion of *acrD* or *acrEF* from *S.* Typhimurium can enhance AcrB expression [91]. The AcrD efflux pump was shown to extrude and capture aminoglycosides along with its role in the efflux of taurocholate. AcrD actively extrudes kanamycin, tobramycin, gentamicin, and amikacin producing an increased transport of protons across the membrane (Table 1) [48]. The *acrD* or *mdtABC* genetic elements encode pumps that confer resistance against novobiocin, SDS and deoxycholate in *S. enterica* cells [93]. Mutations in *acrR* de-repress expression of *acrB*, thereby activating the drug efflux pump function of the AcrAB system [91].

The AcrEF Efflux Pump

In *Salmonella,* AcrEF from the RND family was found to be homologous with AcrAB from the same transporter family [94, 95]. The AcrEF efflux pump is known to functionally complement the AcrAB–TolC system [95]. The AcrD or AcrEF efflux pumps have been shown to confer mild resistances to nalidixic acid, ciprofloxacin, tetracycline, chloramphenicol, triclosan and CTAB (Table 1) [91]. Mutants lacking either of the AcrD or AcrEF efflux pumps did not show high susceptibilities to clinically relevant antimicrobials, suggesting their roles in conferring clinical resistances via this active efflux mechanism in *Salmonella* [91, 96]. This finding indicates that a fully intact AcrAB-TolC pump system is minimally required in order to be a major resistance mechanism conferring MDR in most serovars of *S. enterica* [91]. AcrEF has been shown to exhibit a prominent role in cell division and biofilm formation, thereby leading to an enhancement of virulence [94]. The *acrEF* genes encoding the AcrEF proteins conferred multidrug resistance in a fashion similar to that of AcrAB [93]. AcrEF and EmrAB were involved in the efflux of triclosan from the cell and complemented the role of AcrAB when over-expressed [95].

AcrEF is repressed by the local repressor AcrS. Over-expression of AcrEF is regulated by the presence of IS transposable elements (IS-1 and -10) upstream of the gene encoding AcrS [91, 96]. These IS-1 and -10 elements confer fluoroquinolone resistance against marbofloxacin, enrofloxacin, ciprofloxacin, florfenicol and erythromycin in *Salmonella* [96]. From previous studies it is known that in a few strains, like *S.* Typhimurium DT104, IS-1 and -10 mediate the MDR phenotype by enhancing the expression of the acrEF genes [91].

The MdtABC Efflux System

Of the five putative drug efflux pump systems annotated from the RND family, the transport abilities and roles in drug resistance and virulence of the MdtABC and MdsABC pumps (below) are poorly understood. The predicted MdtB and MdtC proteins are encoded on two different genomes but they function as one transporter system to actively efflux antibiotics [93]. The genetic elements *acrD* or *mdtABC* encoding pumps confer resistance against novobiocin, SDS and DOC (Table 1) [93]. MdtABC from *Salmonella* possesses two different RND transporters, MdtB and MdtC, both of which are co-transcribed in an operon along with a membrane fusion protein, MdtA [97]. The functional pump consists of a hetero-multimeric unit formed by two subunits of MdtB and one subunit of MdtC [97]. MdtC is involved in the binding and transport of drugs, and MdtB induces a conformational change needed for transport using proton translocation (proton motive force) as an energy driving force. In *S. enterica*, MdtABC confers resistance to β-lactams, novobiocin, SDS, and bile salts, and it detoxifies the cell from zinc, copper and tungstate (Table 1) [98].

The MdsABC Efflux Pump System

The *mdsABC* system of *S. enterica* is absent in other bacterial species and is at present surprisingly unique to the *Salmonella* genome. The *mdsA* gene encodes an outer membrane protein whereas MsdB is an RND efflux pump [47]. MdsC of *Salmonella*, like TolC, is highly expressed in drug resistant variants of *Salmonella*, owing to their major role in active drug efflux [47]. The RND-type MdsAB system could function with either MdsC, which is encoded in the same operon, or TolC as the outer membrane component [93]. MsbA from *S.* Typhimurium has two ATP binding cassette (ABC) domains (see below) and nucleotide binding domains for the binding and hydrolysis of ATP [99]. The *Salmonella*-specific MdsABC system confers resistance to a variety of toxic compounds [93]. Overproduction of the MdsABC transporter confers bacterial resistance to novobiocin, acriflavine, crystal violet, ethidium bromide, methylene blue, rhodamine 6G, tetraphenylphosphonium bromide (TPP), benzalkonium bromide and SDS (Table 1) [93].

The ABC Transporter Family of Drug Efflux Pumps

The primary active multidrug ATP-binding cassette (ABC) transporters play an important role in the extrusion of drugs from the bacterial cell, as well the efflux of anticancer drugs from eukaryotic cells [100]. The ABC solute transport systems utilize the hydrolysis of ATP as the energy driving force; and these pumps contain two nucleotide-binding domains (NBDs) where ATP is bound and hydrolyzed and two membrane domains (MDs) which mediate transport of substrates across the cell membrane.

The MacAB-TolC Efflux Pump System

The MacAB-TolC drug transport system from *S.* Typhimurium belongs to the ABC family of solute transporters [47]. MacAB-TolC is an ABC-type efflux complex consisting of the periplasmic protein, MacA, an efflux pump, MacB, and TolC, an outer membrane channel protein [80, 101]. Of the efflux pumps from *Salmonella*, the macrolide-specific pump MacAB, is one of the few ABC-type drug efflux pumps from *Salmonella* that have been studied in detail; and this transporter is a major contributing factor to virulence [47, 101, 102]. The *Salmonella* MacAB system can recognize macrolides as their substrates and operates by using the energy obtained from ATP hydrolysis [47]. The energy obtained from ATP hydrolysis is used to actively pump the antibiotics out of cell against their concentration gradient. The efflux of macrolides confers resistance upon the bacterium to these antimicrobial agents. The MacAB drug efflux pump, however, has functions beyond resistance to macrolides and erythromycin. For instance, the MacAB system was proposed to have a specific role in detoxifying natural compounds encountered during infection [93]. The transporter system also plays a role in the protection of *Salmonella* against oxidative stress and is induced upon exposure to hydrogen peroxide, H_2O_2 (Table 1) [93, 102]. MacAB is apparently also required for intracellular multiplication [102]. MacAB mutants with a deletion of the RND-type AcrAB efflux pump (ΔacrAB) still conferred macrolide resistance to *Salmonella* strains [101].

In *Salmonella*, PhoPQ, a two-component signal transduction system, regulates the expression of the MacAB pump directly [47, 80]. The *macAB* drug efflux genes are repressed by PhoP [93]. The PhoP protein binds to a site in the *macA* gene, the intergenic region between *somA* and the *macAB* ABC-transporter genes, and controls the expression of *macAB* [47]. Interestingly, *somA* is a PhoP-activated gene that has been implicated in virulence, and the *macAB* operon is a PhoP-repressed locus. Thus, during infection of a mammalian host the *macAB* genes mediate *Salmonella* activity outside of host cells [93].

THE MATE FAMILY OF DRUG EFFLUX PUMPS

The MATE (Multidrug And Toxic Compound Extrusion) family is the most recently categorized one among the five multidrug efflux transporter families [103]. MdtK from *S. Typhimurium* belongs to this family.

THE MDTK EFFLUX PUMP

MdtK from *S.* Typhimurium is a MATE-type efflux pump. The integral membrane protein pump transports norfloxacin, doxorubicin and acriflavine out of the cell through biological membrane (Table 1), thus conferring bacterial resistance to these antimicrobial agents [103]. *Salmonella* MdtK (YdhE) does not confer TPP and DOC resistance like that observed in *E. coli* [93].

THE TOLC OUTER MEMBRANE PROTEIN

The AcrAB, AcrD, AcrEF and MdtABC transporters described above require the TolC outer membrane channel, as well, in order to completely extrude drugs to the extracellular milieu [90]. Except for the *acrD* gene, all RND efflux system genes also encode an MFP. Thus, the AcrEF, AcrD, MdtABC and EmrAB systems rely on TolC as their outer membrane component for their drug extrusion functions. Deletion of TolC inactivates all four of these efflux systems in *Salmonella* [93].

THE SMALL MULTIDRUG RESISTANCE (SMR) FAMILY OF EFFLUX PUMPS

The SMR family of efflux pumps are secondary active multidrug transporters energized by the proton gradient [104-109] involving an exchange between the drug and a proton across the membrane, translocation of drugs by changes in the pump conformation, followed by replacement of the drug by a proton in a mechanism known as antiport [110]. The SMR proteins are small integral homo-oligomeric membrane proteins consisting of 100-140 amino acids [104, 108, 111] with a molecular weight of approximately 12 kDa [112]. This type of efflux pump is associated with resistances to a variety of quaternary ammonium compounds and lipophilic cations such as methyl viologen, tetraphenylphosphonium (TPP), benzalkonium, cetyltrimethylammonium bromide (CTAB), cetylpyridinium chloride (CTPC), and dyes such as ethidium bromide, acriflavine, proflavin, crystal violet, pyronine Y, and safranin O. The SMR efflux pump also confers resistance to lipophilic anions like sodium dodecyl sulfate (SDS), the antimicrobial agent phosphonomycin, and a neutral drug like chloramphenicol. Resistances to antibiotics such as β-lactams, cephalosporins, dihydrofolate inhibitors and aminoglycosides are also observed [104].

The SMR efflux pump family was first observed in *S. aureus* and other Staphylococci as a *qacC*/*qacD*/*ebr* but later was renamed *smr* [113]. The pump consists of 4 transmembrane proteins [107, 114-119] that are included in a drug/metabolite transporter (DMT) superfamily [104, 112, 113, 120, 121]. In the membrane, an intact SMR pump forms a homotrimer while being soluble in organic solvents due to its hydrophobic nature [122] and is the efflux transporter family that is found only in bacteria [110, 122, 123]. Protein databases using EmrE of *E. coli* as a seed sequence shows that their pump diversity in bacteria is a result of *smr* gene duplication since 52% of completely sequenced bacterial genomes possess SMR homologues [104].

As of this writing, there are approximately 100 SMR protein sequences, and its classification is based on phenotypic characteristics [104]. The SMR transporter family has been divided into three subclasses; (i) the small multidrug pumps (SMP), (ii) the suppressor of *groEL* mutation proteins (SUG) and (iii) the paired small multidrug resistance proteins (PSMR) [104, 124]. These proteins confer multidrug resistance in both Gram-positive and -negative bacteria, plus in the archaea [104].

Both *qacE* and *qacE*Δ*1* are antibiotic resistant genes of the SMP class that are found on an integron in Gram-negative bacteria [113, 123]. The *qacE* gene was originally observed on the 3' conserved segment of class I integron, found in a plasmid from *Klebsiella aerogenes* [113]. The *qacE*Δ*1* gene, encoding a protein of 115 amino acids, is found on the *Salmonella enterica* genome and confers resistance to quaternary ammonium compounds and intercalating dyes [104, 125, 126]. *qacE*Δ*1* is a defective form of *qacE*, being deficient in the last 16 amino acids from the C- terminus of the protein [104, 125, 127].

QacF, which is also located on the class I integron, is 68% similar to QacE. QacF is found in *S. enterica* YP_209355 with a length of 110 amino acids [104]. In one study, QacC, an SMR pump from *S. epidermidis* was shown to confer high level of resistances to oxacillin, cefazoline and cephaloridine but to a lesser extent to cephadroxil, ceftazidine and ethidium bromide when expressed in Gram-negative hosts like *S.* Typhimurium; but the same pump confers low level resistance to these antibiotics when expressed in its natural host [117].

SUG is also called SugE, and its classification is based also upon phenotypic characteristics [104, 128]. In *Salmonella* ssp., SugE is an SMR protein [129] with 105 amino acids [104]. Rather than transporting quaternary ammonium compounds and lipophilic dyes, the SugE protein imports and exports these substrates in small amounts if overexpressed [104]. GroEL is a chaperone that is involved in the folding of proteins [104, 130]. SugE was so named due to its ability to the restore lost nitrogenase activity when GroEL was defective [131], and SugE over expression results in the suppression of the GroEL mutation [104, 132].

The third subclass, PSMR, requires two separate SMR genes to be co-expressed in order to confer drug resistance. This protein consists of longer hydrophilic loops and an extended hydrophilic C-terminus making one of the SMR proteins longer than the other one, e.g., in the YdgEF system, the YdgF protein contains 121 amino acids; while YdgE has 109 residues. There is single PSMR protein pair in *E. coli* whereas in *B. subtilis* it consists of 3 or more pairs, suggesting that number of protein pairs depends on the organism [104].

The Major Facilitator Superfamily

The major facilitator superfamily (MFS) consists of a very large number of solute transporters as members [133]. These transporters have related amino acid sequences, similar predicted numbers of transmembrane domains, and are thought to share a common evolutionary ancestor [134]. The transporters of the MFS may have as their main modes of energy both secondary active and passive mechanisms [135]. Amon the secondary active transporters, members of the MFS include both symporters and antiporters [136, 137], and among the passive transporters, MFS members include uniporters [133]. Although the MFS share similarities in sequence, structure, origin and modes of energy, these transporters nonetheless harbor extreme differences in their substrates, such as in the number of substrates and substrate structure [85, 135]. With respect to the MFS transporters of *S. enterica*, many are believed to consist of a variety of drug and multidrug efflux pumps.

The MFS Antimicrobial Efflux Pumps

The genome sequencing of bacterial pathogens is a rapidly expanding field [138, 139], and those genome projects pertaining to *S. enterica* have helped to identify virulence factors associated with this pathogenic organism [140-143]. Towards this, systematic evaluation of *S. enterica* genome sequences has resulted in the discovery of determinants that are predicted to encode drug resistance mechanisms [68, 144]. The major facilitator superfamily represents one of the largest transporter superfamilies, consisting of uniporters, symporters and antiporters with 12- or 14-transmembrane domains and similar amino acid sequences but with diverse substrates, such as ions, sugars, amino acids, metabolic intermediates and antimicrobial agents [85, 133, 145-151].

The EmrB and MdfA Multidrug Efflux Pumps of the MFS

Although the MFS represents one of the largest transporter superfamilies, with thousands of members [135], these transporters are relatively poorly studied in *S. enterica*. The *S. enterica* multidrug efflux pumps EmrB [152, 153] and MdfA [86, 93] are so far the only well characterized members of the MSF. The EmrB transporter, an integral inner-membrane protein, is associated with a periplasmic protein, EmrA, and an outer-membrane channel-like protein, TolC, to form a tripartite complex collectively conferring resistance to a several antimicrobial agents, like novobiocin and nalidixic acid in *S. enterica* [86, 95]. The MdfA antimicrobial efflux pump from *S. enterica* is also a member of the MFS. The *mdfA* gene from *Salmonella* encoding the MdfA drug pump confers resistance to chloramphenicol doxorubicin, norfloxacin, and tetracycline, but not to acriflavine, ethidium bromide and tetraphenylphosphonium (TPP) [93]. However, MdfA from *E. coli* actively transports these antimicrobial agents [93]. The reason for these observed differences in the substrate profiles remains to be elucidated. The over-expression of MdfA in *Salmonella* increases alkaline tolerance by enhancing an exchange process with $Na^+(K^+)$ / H^+ in addition to mediating antimicrobial drug / H^+ antiport [154].

Using MdfA from *S. enterica* as a query sequence, we conducted a BLASTP analysis and found that the *S. enterica* chromosome surprisingly harbors about 16 uncharacterized putative drug efflux pumps, all belonging to the MFS. These putative and so far uncharacterized drug efflux pumps from *S. enterica* are most closely related to MdtM, SmvA, MdtH, Tcr4, MdtD, MdtM-2, SalT1, SaliN, MdtM-3, SaleT, MdtM-4, MdfA-1, MdfA-2, MdfA-3, Cmr, and MdtH from a variety of other microorganisms. Clearly, much molecular and physiological work must be completed before these putative determinants can be considered MFS drug and multidrug efflux pumps of *S. enterica*. It is thus important that these genomic elements be studied in order to find means to circumvent these potential antimicrobial resistance mechanisms. Such studies would potentially restore the clinical efficacy of certain antimicrobials and facilitate treatment efforts against salmonellosis.

THE *SALMONELLA* GENOMIC ISLAND I (SGI-1)

The *Salmonella* pathogenicity island 1 (SPI-1) harbors genes encoding factors necessary for the invasion of intestinal epithelial cells and the induction of intestinal secretory and inflammatory responses [155]. On the other hand, the *Salmonella* pathogenicity island 2 (SPI-2) encodes genes that constitute a type III secretion system for the translocation of virulence factors to target host cells [155]. The 43-Kbp size SGI-1 element from *S.* Typhimurium confers multidrug resistance to ampicillin (A), chloramphenicol (C), streptomycin (S), sulfonamides (Su) and tetracycline's (T), denoted as ACSSuT [49]. In S. Typhimurium DT104, the MDR conferring genes are localized on SGI1 and mediate resistance to ampicillin (*bla*-$_{PSE}$), chloramphenicol and florfenicol (*floR / cmlA*), gentamicin (*aac(3)*), streptomycin and spectinomycin (*aadA* and *aadA7*), sulfonamides (*sul1*), tetracyclines (*tetA(G)*), and trimethoprim (*dfrA1* and *dfrA10*) [49, 155]. Integrons are capable of transferring antibiotic resistance determinants, and the resistance encoding locus in SGI (13-Kbp) is flanked by two class 1 integrons [49]. This region contains an *aadA2* gene (conferring resistance to streptomycin and spectinomycin) on the proximal side and the β-lactamase gene, *bla*-$_{PSE}$, on the distal end. In SGI integrase enzyme encoded by the *intI1* gene, located at 5’- conserved segment of class-1 integrons, facilitates recombination, expression and recognition of recombined *attI* [49]. The 3’- conserved segment of SGI encodes *qacE1* and *sul1*, which confer resistance to quaternary ammonium compounds and sulfonamides, respectively [49]. From studies by Abraham et al., *S. enterica* strains exhibiting the MDR phenotype isolated from Australian livestock harbored *bla*$_{TEM}$, *sul2, tetB, tetA, aphA1 , dhfrV, sul1, aadA, aac(3)-IV, dhfrI, tetC,* and *cmlA* genes associated with class 1 integrons (*intI*) [156]. The *aadB, aphAI-IAB, aadA1, aadA2,* or *bla*$_{PSE1}$, *bla*$_{TEM}$, *cat1, cat2, dhfr1, floR, strA, sul1, sul2, tetA(A), tetA(B)* and *tetA(G)* genes are the resistance conferring clusters presently integrated in the class 1 integrons [157]. The *aadA1* and *aadA2* genes confer resistance to spectinomycin and streptomycin by encoding the enzyme spectinomycin / streptomycin adenyl tranferase. The *aadB* gene encodes an aminoglycoside transferase enzyme that inactivates gentamicin. The *aphA1-1AB* determinant encodes an aminoglycoside phosphotransferase which confers resistance to kanamycin. The bla_{CARB2} and bla_{TEM} genes encode β-lactamases which degrade ampicillin. CAT1 and -2 are chloramphenicol acetyl-transferases that protect the cell from chloramphenicol. DHFR1 is a dihydrofolate reductase enzyme that can degrade trimethoprim. StrA is a streptomycin phosphotransferase and *sul1* and *2* encode dihydropteroate synthases that protect from spectinomycin and sulfadiazine (Table 2). The *blaP1* element found on the class-1 integron codes for PSE-1 / CARB-2 and is present on SGI1 [157]. A new gene *cmY-2* coding a β-lactamase is effective in inactivating third-generation cephalosporins leading to resistances against ampicillin, amoxicillin-clavulanic acid, cefoxitin, cephalothin, ceftiofur, chloramphenicol, streptomycin, sulfamethoxazole, and tetracycline (Figure 2) [155, 156, 158].

Table 2. Resistance-conferring genes present on SGI-1

Gene	Protein	Substrates
aadA2	β-lactamase gene	Streptomycin and spectinomycin
int-III	SGI Integrase	Integration
aadA1 and *aadA2*	Spectinomycin/ streptomycin adenyl transferase	Spectinomycin, streptomycin
aadB	Aminoglycoside transferase	Gentamicin
aphA1-1AB	Aminoglycoside phosphotransferase	Kanamycin
bla_{TEM}	β-lactamases	Ampicillin
cat1 and *2*	Chloramphenicol acetyl-transferases	Chloramphenicol
dhfr1	Dihydrofolate reductase	Trimethoprim
strA	Streptomycin phosphotransferase	Streptomycin and spectinomycin
sul1 and *2*	Dihydropteroate synthases	Sulfadiazine
blaP1/ pse-1/ carb-2	β-lactamase	Ampicillin
aac(3)	AAC(3)-I aminoglycoside 3-*N*-acetyltransferase	Gentamicin
cmY-2	β-lactamase	Third-generation cephalosporins
bla_{SHV-12}	β-lactamase	Sulfamethoxazole, chloramphenicol, ampicillin, tetracycline, kanamycin, trimethoprim, amoxicillin-clavulanic acid, gentamicin, streptomycin, and extended-spectrum cephalosporins
bla_{TEM-1b}	β-lactamase	Sulfamethoxazole, chloramphenicol, ampicillin, tetracycline, kanamycin, trimethoprim, amoxicillin-clavulanic acid, gentamicin, streptomycin, and extended-spectrum cephalosporins

The enzymes encoded by the genes of the SGI-1 element and antibiotics that the enzymes degrade are shown.

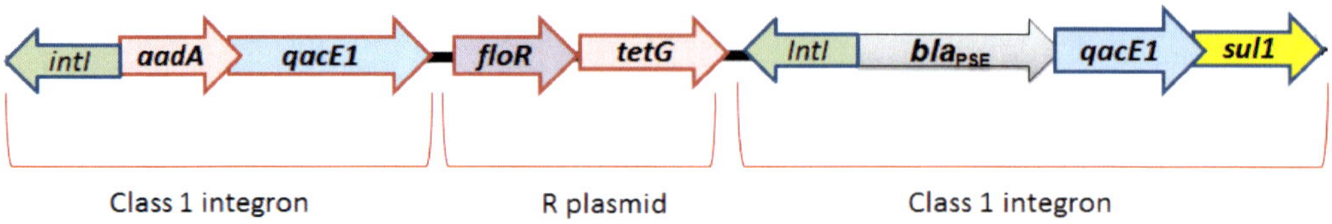

Figure 2. Gene arrangement of the SGI-1 locus in *S. enterica*.

In Spain the ESBL-producing *S. enterica* isolates harbored the *blaSHV-12* gene combined with *bla*$_{TEM}$*-1b, tet(A), aadA* and *sul1* genes (Table 2) [159]. These isolates were resistant to ciprofloxacin and exhibited high-level resistances to sulfamethoxazole, chloramphenicol, ampicillin, tetracycline, kanamycin, trimethoprim, amoxicillin-clavulanic acid, gentamicin, streptomycin, and extended-spectrum cephalosporins [160]. Extremely drug-resistant (XDR) *S*. Typhimurium isolates were also resistant to all three of these choices of treatment, namely, ciprofloxacin, ceftriaxone, and azithromycin [160]. Interaction of drug resistance conferring genes on *S*GI-1 with AcrAB-TolC tripartite efflux pump confer resistance to chloramphenicol, florfenicol and tetracyclines.

THE QRDR SYSTEM

The quinolone resistance phenotype observed in *Salmonella* was first attributed to the presence of mutations in the quinolone resistance determining region (QRDR) of the gyrase or in topoisomerase proteins [92]. Mutations occurring in the gyrase QRDR region (*gyrA* and *gyrB*) and topoisomerase IV (*parC* and *parE*) were implicated as significant determinants of fluoroquinolone resistance in *S.* Typhimurium [49]**.** In the QRDR between amino acids 67 and 106, specifically at residues Ser-83 and Asp-87, an association was found with resistance to nalidixic acid and many fluoroquinolones [92]. Work by Giruad *et. al* demonstrated that *S.* Typhimurium mutants lacking alterations either in gyrase or in topoisomerase IV were demonstrated to exhibit ciprofloxacin resistance due to the AcrAB active efflux acting alone, thus speculating that QRDR and active efflux mediate a synergistic outcome [161].

TRANSCRIPTIONAL ACTIVATORS, REGULATORS AND INDUCERS

Expression of stable efflux pumps within the biological membrane of a bacterium is tightly regulated by many transcriptional factors [80]. The genes *acrR*, *marRAB*, *soxRS* and *ramR* are MDR regulatory elements in *Salmonella* [152]. Expression from the *acrAB* genetic element is modulated locally by the repressor AcrR and at a global level by master regulators such as MarA, SoxS, and Rob [162, 163].

THE MARRAB LOCUS

The *marR* and *marA* genes encode a transcriptional repressor and an activator, respectively, both of which regulate the development of the MDR phenotype in *S. enterica* by drastically reducing the entry and increasing the active efflux of antibiotics via AcrAB efflux pump activity [164]. The *marRAB*, *acrAB* and *tolC* genes are positively regulated by MarA and SoxS in *Salmonella spp.* (Table 3) [165]. The *marRAB* locus is modulated by a variety of factors, including organic compounds [166] plus milk-containing foods [167] and milk itself [168].

THE SOXS TRANSCRIPTIONAL ACTIVATOR

SoxS, an effector of the *soxRS* global superoxide response regulon was demonstrated to transcribe *micF* and *acrAB* by binding to marbox, a binding site for SoxS, thereby activating the MDR phenotype in *S. enterica* (Table 3) [165]. Importantly, increased expression of *marA* and *soxS* at the *marRAB* locus results in an up-regulation of efflux pump systems like *acrAB*, thus conferring the MDR phenotype in *Salmonella* strains [49].

THE RAMA TRANSCRIPTIONAL ACTIVATOR

RamA is a regulatory protein from the AraC-XylS transcriptional activator family, identified in *S. enterica* serovars Paratyphi-B and Typhimurium conferring high-level resistance to chloramphenicol, tetracycline, tigecycline, fluoroquinolones, and trimethoprim [169]. RamA has also been demonstrated to activate the *marRAB* locus [164]. RamA and SoxS together play a role in the oxidative stress response by *Salmonella* spp. [164]. The presence of indole and bile, fluids present in the internal human environment, like the gastrointestinal track, activate the AcrAB pump by inducing RamA (Table 3). Thus far, RamA is unique to *Salmonella* strains, *Klebsiella pneumoniae* and *Enterobacter* species but is apparently absent in other bacterial populations [47]. Studies also indicate that upon inactivation of the *ramR* gene, the repressor for *ramA*, results in an expression of the MDR phenotype by increasing the expression from *ramA* and the production of the AcrAB efflux pump [47].

THE BAESR-TCS REGULATORY SYSTEM

The two-component regulatory system BaeSR increases multidrug and metal resistances in *Salmonella* through the induction of drug efflux systems. The *baeR* gene encodes the response regulator of BaeSR [163]. The BaeSR system acts as an activator for the expression of the *mdtABC* operon, *acrD*, and *tolC* in response to indole, copper, or zinc [47, 163]. In *Salmonella* strains the presence of copper or zinc in the media was shown to activate the BaeSR system thereby inducing the expression of *mdtABC* and *acrD* determinants to confer resistance against β-lactams, novobiocin and deoxycholate (Table 3) [163]. Over-expression of *baeR* results in an increase in the resistances to oxacillin, cloxacillin, nafcillin, novobiocin, deoxycholate, and β-lactams in *Salmonella* [47, 163]. BaeSR, AcrD, and MdtABC contribute to copper and zinc resistance in *Salmonella*. BaeR also activates the expression of *mdtD*, which encodes a putative MFS-type transporter that is so far not related to the MDR phenotype [163].

THE PHO-PQ TCS MASTER RESPONSE REGULATOR

The *phoP* / *phoQ* genetic element specifies a response regulator, a two-component virulence conferring system. PhoP activates the *pag* gene required for intra-macrophage survival and resistance against antibiotics [93]. Pho-PQ TCS is a master regulator of virulence in *Salmonella* and can sense and respond to low magnesium levels and low pH conditions [80]. Mutants lacking this regular become susceptible to antibiotics and an acidic pH [93].

Table 3. Transcriptional inducers, repressors and regulators of different efflux pump genes in *S. enterica*

Gene(s)	Regulators	Regulator Family	Inducer	Repressors
acrAB	RamR	TetR	Indole, Bile	AcrA
	AcrR	TetR		
	MarA	AraC		MarR
	SoxS	AraC		
	BaeS-BaeR	Two- component system BaeSR	Indole, zinc, copper	
acrD	CpxA-CpxR	Two- component system BaeSR	Indole, zinc, copper	
	BaeS-BaeR	Two- component system BaeSR		
mdtABC	CpxA-CpxR	Two- component system BaeSR		
	GolS	MerR		
mdsABC	AcrS	TetR		
acrEF	PhoQ-PhoP	TCS-Pho-PQ		
macAB	PhoQ-PhoP	TCS-Pho-PQ		
emrAB	PhoQ-PhoP	TCS-Pho-PQ		EmrR
mdfA	Unknown	Unknown		
mdtK	Unknown	Unknown		

FUTURE DIRECTIONS

Infections in which the causative agents are drug resistant *S. enterica* microorganisms represent a serious public health concern. Efforts to facilitate conditions that reduce incidence, prevalence, and dissemination through populations are clearly desirable. Appropriate, responsible, and prudent use of antimicrobials, i.e., antimicrobial stewardship, in medicine and agriculture may slow the onset of drug and multidrug resistant variants of *S. enterica* [170-173]. One clear avenue is the development of new antibacterial agents for therapy. Novel antimicrobials, however, are either non-existent or are underdeveloped, and new drugs may require a long period of time before their demonstration of utility, safety, and regulatory approval allow them to come to fruition. As drug and multidrug efflux pumps of *S. enterica* confer resistance to clinical antimicrobial agents and potentially reduce their clinical efficacy, these transporters make good targets for the development of novel efflux pump inhibitors [174, 175]. Modulation of the systems that regulate efflux pump gene expression may be considered as a promising avenue for restoration of the efficacy of antimicrobial agents to which *S. enterica* clinical isolates are recalcitrant [176]. Combination therapy in which two structurally distinct drugs are more effective together than either of the single drugs alone, i.e., synergy, may be considered, as long as treatment regimens are completed [177]. Phage therapy, in which bacteriophage virus specific to *S. enterica* could be utilized for treatment of salmonella infections, may be considered, as such phage treatments are available for listeriosis caused *by Listeria monocytogenes* [178].

ACKNOWLEDGMENTS

This work was supported in part by a grant from the National Institutes of Health (NIH P20GM103451) and an Internal Research Grant (IRG) from ENMU.

REFERENCES

[1] Su L, Chiu C: Salmonella: clinical importance and evolution of nomenclature. *Chang Gung medical journal* 2007, 30:210.

[2] Murray B, Jorgensen L: Pfaller (ed.). 2007. *Manual of clinical microbiology, 9th ed American Society for Microbiology, Washington, DC.*

[3] Jones TF, Ingram LA, Cieslak PR, Vugia DJ, Tobin-D'Angelo M, Hurd S, Medus C, Cronquist A, Angulo FJ: Salmonellosis outcomes differ substantially by serotype. *Journal of Infectious Diseases* 2008, 198:109-114.

[4] Kingsley RA, Bäumler AJ: Host adaptation and the emergence of infectious disease: the Salmonella paradigm. *Molecular microbiology* 2000, 36:1006-1014.

[5] Foley S, Lynne A, Nayak R: Salmonella challenges: prevalence in swine and poultry and potential pathogenicity of such isolates. *Journal of animal science* 2008, 86:E149-E162.

[6] Callaway T, Edrington T, Anderson R, Byrd J, Nisbet D: Gastrointestinal microbial ecology and the safety of our food supply as related to Salmonella. *Journal of Animal Science* 2008, 86:E163-E172.

[7] Foltz V: Salmonella ecology. *Journal of the American Oil Chemists' Society* 1969, 46:222-224.

[8] Threlfall E, Frost J, Ward L, Rowe B: Increasing spectrum of resistance in multiresistant Salmonella typhimurium. *The Lancet* 1996, 347:1053-1054.

[9] Scallan E, Hoekstra RM, Angulo FJ, Tauxe RV, Widdowson M-A, Roy SL, Jones JL, Griffin PM: Foodborne illness acquired in the United States—major pathogens. *Emerg Infect Dis* 2011, 17.

[10] Ryan KJ, Ray CG: *Sherris medical microbiology.* McGraw Hill Medical New York; 2010.

[11] Murray PR, Rosenthal KS, Pfaller MA: *Medical Microbiology 7 edition.* Elsevier Health Sciences; 2013.

[12] Buxton A, Fraser G: *Animal microbiology. Volume 1: immunology, bacteriology, mycology, diseases of fish and laboratory methods.* Blackwell Scientific Publications.; 1977.

[13] Rice DH, Besser TE, Hancock DD: Epidemiology and virulence assessment of Salmonella dublin. *Veterinary microbiology* 1997, 56:111-124.

[14] Berlanga M: Brock Biology of Microorganisms (11th edn). Michael T. Madigan, John M. Martinko (eds). *International Microbiology* 2010, 8:149-150.

[15] Pardon P, Sanchis R, Marly J, Lantier F, Pepin M, Popoff M: [Ovine salmonellosis caused by Salmonella abortus ovis]. *Annales de recherches veterinaires Annals of veterinary research* 1987, 19:221-235.

[16] Schwartz K: Salmonellosis in swine. *The Compendium on continuing education for the practicing veterinarian (USA)* 1991.

[17] Pang T, Bhutta ZA, Finlay BB, Altwegg M: Typhoid fever and other salmonellosis: a continuing challenge. *Trends in microbiology* 1995, 3:253-255.

[18] Connor BA, Schwartz E: Typhoid and paratyphoid fever in travellers. *The Lancet infectious diseases* 2005, 5:623-628.

[19] Bronze MS, Greenfield RA: *Biodefense: principles and pathogens.* Horizon Bioscience; 2005.

[20] Tsolis RM, Young GM, Solnick JV, Bäumler AJ: From bench to bedside: stealth of enteroinvasive pathogens. *Nature Reviews Microbiology* 2008, 6:883-892.

[21] Sechter I, Katzenelson E, Reisfeld A: Salmonella serovars (others than Typhi and Paratyphi) from extra-intestinal sources. Israel, 1984–9. *Epidemiology and infection* 1991, 106:485-488.

[22] Mills SD, Finlay BB: Comparison of Salmonella typhi and Salmonella typhimurium invasion, intracellular growth and localization in cultured human epithelial cells. *Microbial pathogenesis* 1994, 17:409-423.

[23] Zaidi E, Bachur R, Harper M: Non-typhi Salmonella bacteremia in children. *The Pediatric infectious disease journal* 1999, 18:1073-1077.

[24] Chen Y-H, Chen T-P, Lu P-L, Su Y-C, Hwang K-P, Tsai J-J, Cheng H-H, Peng C-F: Salmonella choleraesuis bacteremia in southern Taiwan. *The Kaohsiung journal of medical sciences* 1999, 15:202-208.

[25] Chiu C, Lin T, Ou J: Predictors for extraintestinal infection of non-typhoidal Salmonella in patients without AIDS. *International journal of clinical practice* 1998, 53:161-164.

[26] Cohen JI, Bartlett JA, Corey GR: Extra-intestinal manifestations of salmonella infections. *Medicine* 1987, 66:349-388.

[27] Sirinavin S, Jayanetra P, Lolekha S, Layangkul T: Predictors for extraintestinal infection in Salmonella enteritis in Thailand. *The Pediatric infectious disease journal* 1988, 7:44-47.

[28] Chiu C-H, Lin T-Y, Ou JT: Age-related differences of nontyphoid Salmonella bacteremia in clinical presentation and outcome: association with specific serovars but not necessarily with the virulence plasmids. *Clinical Infectious Diseases* 2000, 30:239-240.

[29] Chiu CH, Lin TY, Ou JT: Prevalence of the virulence plasmids of nontyphoid Salmonella in the serovars isolated from humans and their association with bacteremia. *Microbiology and immunology* 1999, 43:899-903.

[30] Threlfall E, Hall M, Rowe B: Salmonella bacteraemia in England and Wales, 1981-1990. *Journal of clinical pathology* 1992, 45:34-36.
[31] Osler W: The Gulstonian lectures, on malignant endocarditis. *British medical journal* 1885, 1:522.
[32] Oskoui R, Davis WA, Gomes MN: Salmonella aortitis: a report of a successfully treated case with a comprehensive review of the literature. *Archives of internal medicine* 1993, 153:517-525.
[33] Barrow P, Huggins M, Lovell M, Simpson J: Observations on the pathogenesis of experimental Salmonella typhimurium infection in chickens. *Research in veterinary science* 1987, 42:194-199.
[34] Kingsley RA, van Amsterdam K, Kramer N, Bäumler AJ: The shdA Gene Is Restricted to Serotypes ofSalmonella enterica Subspecies I and Contributes to Efficient and Prolonged Fecal Shedding. *Infection and immunity* 2000, 68:2720-2727.
[35] Kingsley RA, Humphries AD, Weening EH, de Zoete MR, Winter S, Papaconstantinopoulou A, Dougan G, Bäumler AJ: Molecular and phenotypic analysis of the CS54 island of Salmonella enterica serotype Typhimurium: identification of intestinal colonization and persistence determinants. *Infection and immunity* 2003, 71:629-640.
[36] Crawforda RW, Rosales-Reyesb R, de la Luz Ramírez-Aguilarb M, Chapa-Azuelac O, Alpuche-Arandad C, Gunna JS: Gallstones play a significant role in Salmonella spp. gallbladder colonization and carriage. *PNAS* 2010, 107:4353-4358.
[37] Sanchez-Vargas FM, Abu-El-Haija MA, Gomez-Duarte OG: Salmonella infections: an update on epidemiology, management, and prevention. *Travel Med Infect Dis* 2011, 9:263-277.
[38] Swaminathan B, Barrett TJ, Fields P: Surveillance for human Salmonella infections in the United States. *J AOAC Int* 2006, 89:553-559.
[39] Gunn JS, Marshall JM, Baker S, Dongol S, Charles RC, Ryan ET: Salmonella chronic carriage: epidemiology, diagnosis, and gallbladder persistence. *Trends Microbiol* 2014.
[40] Voetsch AC, Van Gilder TJ, Angulo FJ, Farley MM, Shallow S, Marcus R, Cieslak PR, Deneen VC, Tauxe RV, Emerging Infections Program FoodNet Working G: FoodNet estimate of the burden of illness caused by nontyphoidal Salmonella infections in the United States. *Clin Infect Dis* 2004, 38 Suppl 3:S127-134.
[41] Majowicz SE, Musto J, Scallan E, Angulo FJ, Kirk M, O'Brien SJ, Jones TF, Fazil A, Hoekstra RM, International Collaboration on Enteric Disease 'Burden of Illness S: The global burden of nontyphoidal Salmonella gastroenteritis. *Clin Infect Dis* 2010, 50:882-889.
[42] Uzzau S, Brown DJ, Wallis T, Rubino S, Leori G, Bernard S, Casadesus J, Platt DJ, Olsen JE: Host adapted serotypes of Salmonella enterica. *Epidemiol Infect* 2000, 125:229-255.
[43] Wattiau P, Boland C, Bertrand S: Methodologies for Salmonella enterica subsp. enterica subtyping: gold standards and alternatives. *Appl Environ Microbiol* 2011, 77:7877-7885.
[44] Olsen SJ, Bishop R, Brenner FW, Roels TH, Bean N, Tauxe RV, Slutsker L: The changing epidemiology of salmonella: trends in serotypes isolated from humans in the United States, 1987-1997. *J Infect Dis* 2001, 183:753-761.

[45] Crim SM, Iwamoto M, Huang JY, Griffin PM, Gilliss D, Cronquist AB, Cartter M, Tobin-D'Angelo M, Blythe D, Smith K, et al.: Incidence and trends of infection with pathogens transmitted commonly through food--Foodborne Diseases Active Surveillance Network, 10 U.S. sites, 2006-2013. *MMWR Morb Mortal Wkly Rep* 2014, 63:328-332.

[46] Carrique-Mas JJ, Davies RH: Bacteriological detection of Salmonella enteritidis in eggs: a review. *Rev Sci Tech* 2008, 27:657-664.

[47] Nishino K, Nikaido E, Yamaguchi A: Regulation and physiological function of multidrug efflux pumps in Escherichia coli and Salmonella. *Biochim Biophys Acta* 2009, 1794:834-843.

[48] Rensch U, Klein G, Kehrenberg C: Analysis of triclosan-selected Salmonella enterica mutants of eight serovars revealed increased aminoglycoside susceptibility and reduced growth rates. *PLoS One* 2013, 8:e78310.

[49] Quinn T ROM, Alan W. Baird, Denise Drudy, Paul Whyte and, Fanning. S: Multi-Drug Resistance in *Salmonella enterica*: Efflux mechanisms and their relationships with the development of chromosomal resistance gene clusters. *Current Drug Targets* 2006, 7:849-860.

[50] Foster JW, Moreno M: Inducible acid tolerance mechanisms in enteric bacteria. *Novartis Found Symp* 1999, 221:55-69; discussion 70-54.

[51] Baik HS, Bearson S, Dunbar S, Foster JW: The acid tolerance response of Salmonella typhimurium provides protection against organic acids. *Microbiology* 1996, 142 (Pt 11):3195-3200.

[52] Clark MA, Jepson MA, Simmons NL, Hirst BH: Preferential interaction of Salmonella typhimurium with mouse Peyer's patch M cells. *Res Microbiol* 1994, 145:543-552.

[53] Velge P, Wiedemann A, Rosselin M, Abed N, Boumart Z, Chausse AM, Grepinet O, Namdari F, Roche SM, Rossignol A, Virlogeux-Payant I: Multiplicity of Salmonella entry mechanisms, a new paradigm for Salmonella pathogenesis. *Microbiologyopen* 2012, 1:243-258.

[54] Baumler AJ, Tsolis RM, Heffron F: The lpf fimbrial operon mediates adhesion of Salmonella typhimurium to murine Peyer's patches. *Proc Natl Acad Sci U S A* 1996, 93:279-283.

[55] Francis CL, Starnbach MN, Falkow S: Morphological and cytoskeletal changes in epithelial cells occur immediately upon interaction with Salmonella typhimurium grown under low-oxygen conditions. *Mol Microbiol* 1992, 6:3077-3087.

[56] Fabrega A, Vila J: Salmonella enterica serovar Typhimurium skills to succeed in the host: virulence and regulation. *Clin Microbiol Rev* 2013, 26:308-341.

[57] Hersh D, Monack DM, Smith MR, Ghori N, Falkow S, Zychlinsky A: The Salmonella invasin SipB induces macrophage apoptosis by binding to caspase-1. *Proc Natl Acad Sci U S A* 1999, 96:2396-2401.

[58] Jiang X, Rossanese OW, Brown NF, Kujat-Choy S, Galan JE, Finlay BB, Brumell JH: The related effector proteins SopD and SopD2 from Salmonella enterica serovar Typhimurium contribute to virulence during systemic infection of mice. *Mol Microbiol* 2004, 54:1186-1198.

[59] Alpuche-Aranda CM, Racoosin EL, Swanson JA, Miller SI: Salmonella stimulate macrophage macropinocytosis and persist within spacious phagosomes. *J Exp Med* 1994, 179:601-608.

[60] Kuhle V, Hensel M: Cellular microbiology of intracellular Salmonella enterica: functions of the type III secretion system encoded by Salmonella pathogenicity island 2. *Cell Mol Life Sci* 2004, 61:2812-2826.

[61] Waterman SR, Holden DW: Functions and effectors of the Salmonella pathogenicity island 2 type III secretion system. *Cell Microbiol* 2003, 5:501-511.

[62] Wistrom J, Norrby SR: Fluoroquinolones and bacterial enteritis, when and for whom? *J Antimicrob Chemother* 1995, 36:23-39.

[63] Stoycheva MV, Murdjeva MA: Antimicrobial therapy of salmonelloses--current state and perspectives. *Folia Med (Plovdiv)* 2006, 48:5-10.

[64] Manchanda V, Bhalla P, Sethi M, Sharma VK: Treatment of enteric fever in children on the basis of current trends of antimicrobial susceptibility of Salmonella enterica serovar typhi and paratyphi A. *Indian J Med Microbiol* 2006, 24:101-106.

[65] Van TT, Nguyen HN, Smooker PM, Coloe PJ: The antibiotic resistance characteristics of non-typhoidal Salmonella enterica isolated from food-producing animals, retail meat and humans in South East Asia. *Int J Food Microbiol* 2012, 154:98-106.

[66] Parry CM, Threlfall EJ: Antimicrobial resistance in typhoidal and nontyphoidal salmonellae. *Curr Opin Infect Dis* 2008, 21:531-538.

[67] Weill FX, Guesnier F, Guibert V, Timinouni M, Demartin M, Polomack L, Grimont PA: Multidrug resistance in Salmonella enterica serotype Typhimurium from humans in France (1993 to 2003). *J Clin Microbiol* 2006, 44:700-708.

[68] Cloeckaert A, Schwarz S: Molecular characterization, spread and evolution of multidrug resistance in Salmonella enterica typhimurium DT104. *Vet Res* 2001, 32:301-310.

[69] Centers for Disease C, Prevention: Outbreak of Salmonella serotype Saintpaul infections associated with multiple raw produce items--United States, 2008. *MMWR Morb Mortal Wkly Rep* 2008, 57:929-934.

[70] Smith KP, George J, Cadle KM, Kumar S, Aragon SJ, Hernandez RL, Jones SE, Floyd JL, Varela MF: Elucidation of Antimicrobial Susceptibility Profiles and Genotyping of Salmonella enterica Isolates from Clinical Cases of Salmonellosis in New Mexico in 2008. *World J Microbiol Biotechnol* 2010, 26:1025-1031.

[71] Higgins CF: Multiple molecular mechanisms for multidrug resistance transporters. *Nature* 2007, 446:749-757.

[72] Kumar S, Varela MF: Molecular mechanisms of bacterial resistance to antimicrobial agents. In *Microbial pathogens and strategies for combating them: science, technology and education.* Edited by Méndez-Vilas A: Formatex Research Center; 2013: 522-534

[73] Kumar S, Floyd JT, He G, Varela MF: Bacterial antimicrobial efflux pumps of the MFS and MATE transporter families: A review. pp. 1-21. In *Recent Research Developments in Antimicrobial Agents & Chemotherapy.* Kerala, India: Research Signpost, Inc.; 2013

[74] Marshall BM, Levy SB: Food animals and antimicrobials: impacts on human health. *Clin Microbiol Rev* 2011, 24:718-733.

[75] Barber DA, Miller GY, McNamara PE: Models of antimicrobial resistance and foodborne illness: examining assumptions and practical applications. *J Food Prot* 2003, 66:700-709.

[76] Nikaido H: Multidrug resistance in bacteria. *Annu Rev Biochem* 2009, 78:119-146.

[77] Alekshun MN, Levy SB: Molecular mechanisms of antibacterial multidrug resistance. *Cell* 2007, 128:1037-1050.
[78] Floyd JT, Kumar S, Mukherjee MM, He G, Varela MF: A review of the molecular mechanisms of drug efflux in pathogenic bacteria: A structure-function perspective. In *Recent Research Developments in Membrane Biology. Volume* 3. Edited by Shankar P. Kerala, India: Research Signpost, Inc.; 2013: 15-66
[79] Floyd JT, Kumar, S., Mukherjee, M.M., He, G.X., and Varela, M.F.: A review of the molecular mechanisms of drug efflux in pathogenic bacteria: A structure-function perspective. *Recent Res Devel Membrane Biol* 2013, 3:15-66.
[80] Sun J, Deng Z, Yan A: Bacterial multidrug efflux pumps: Mechanisms, physiology and pharmacological exploitations. *Biochem Biophys Res Commun* 2014.
[81] Nikaido H, Hall JA: Overview of bacterial ABC transporters. *Methods Enzymol* 1998, 292:3-20.
[82] He G-X, Thorpe C, Walsh D, Crow R, Chen H, Kumar S, Varela M: EmmdR, a new member of the MATE family of multidrug transporters, extrudes quinolones from Enterobacter cloacae. *Arch Microbiol* 2011, 193:759-765.
[83] Nikaido H: Structure and mechanism of RND-type multidrug efflux pumps. *Adv Enzymol Relat Areas Mol Biol* 2011, 77:1-60.
[84] Bay DC, Rommens KL, Turner RJ: Small multidrug resistance proteins: a multidrug transporter family that continues to grow. *Biochim Biophys Acta* 2008, 1778:1814-1838.
[85] Pao SS, Paulsen IT, Saier MH, Jr.: Major facilitator superfamily. *Microbiol Mol Biol Rev* 1998, 62:1-34.
[86] Horiyama T, Yamaguchi A, Nishino K: TolC dependency of multidrug efflux systems in Salmonella enterica serovar Typhimurium. *J Antimicrob Chemother* 2010, 65:1372-1376.
[87] Fuentes DE, Navarro CA, Tantalean JC, Araya MA, Saavedra CP, Perez JM, Calderon IL, Youderian PA, Mora GC, Vasquez CC: The product of the qacC gene of Staphylococcus epidermidis CH mediates resistance to beta-lactam antibiotics in gram-positive and gram-negative bacteria. *Res Microbiol* 2005, 156:472-477.
[88] Nikaido H, Basina M, Nguyen, V, Rosenberg, EY: Multidrug efflux pump AcrAB of Salmonella typhimurium excretes only Those beta-lactam antibiotics containing lipophilic side chains. *Journal of Bacteriology* 1998, 180:4686–4692.
[89] Pos KM: Drug transport mechanism of the AcrB efflux pump. *Biochim Biophys Acta* 2009, 1794:782-793.
[90] Nikaido H, Takatsuka Y: Mechanisms of RND multidrug efflux pumps. *Biochim Biophys Acta* 2009, 1794:769-781.
[91] Piddock LJ: Clinically relevant chromosomally encoded multidrug resistance efflux pumps in bacteria. *Clin Microbiol Rev* 2006, 19:382-402.
[92] Baucheron S, Imberechts, H, Chaslus-dancla E, Cloeckaert A: The AcrB Multidrug Transporter Plays a Major Role in High-Level Fluoroquinolone Resistance in *Salmonella enterica* Serovar Typhimurium Phage Type DT204. *Microbial Drug Resistance* 2002, 8:281-289.
[93] Nishino K, Latifi T, Groisman EA: Virulence and drug resistance roles of multidrug efflux systems of Salmonella enterica serovar Typhimurium. *Mol Microbiol* 2006, 59:126-141.

[94] Baugh S, Ekanayaka AS, Piddock LJ, Webber MA: Loss of or inhibition of all multidrug resistance efflux pumps of Salmonella enterica serovar Typhimurium results in impaired ability to form a biofilm. *J Antimicrob Chemother* 2012, 67:2409-2417.

[95] Rensch U, Nishino K, Klein G, Kehrenberg C: Salmonella enterica serovar Typhimurium multidrug efflux pumps EmrAB and AcrEF support the major efflux system AcrAB in decreased susceptibility to triclosan. *Int J Antimicrob Agents* 2014.

[96] Olliver A, Valle M, Chaslus-Dancla E, Cloeckaert A: Overexpression of the multidrug efflux operon acrEF by insertional activation with IS1 or IS10 elements in Salmonella enterica serovar typhimurium DT204 acrB mutants selected with fluoroquinolones. *Antimicrob Agents Chemother* 2005, 49:289-301.

[97] Guerrero P, Collao B, Alvarez R, Salinas H, Morales EH, Calderon IL, Saavedra CP, Gil F: Salmonella enterica serovar Typhimurium BaeSR two-component system positively regulates sodA in response to ciprofloxacin. *Microbiology* 2013, 159:2049-2057.

[98] Weingart* PDaH: Characterization and regulation of the Resistance-Nodulation-Cell Division-type multidrug efflux pumps MdtABC and MdtUVW from the fire blight pathogen *Erwinia amylovora. BMC Microbiology* 2014, 14:1471-2180.

[99] Seeger MA, van Veen HW: Molecular basis of multidrug transport by ABC transporters. *Biochim Biophys Acta* 2009, 1794:725-737.

[100] Davidson AL, Maloney PC: ABC transporters: how small machines do a big job. *Trends Microbiol* 2007, 15:448-455.

[101] Zgurskaya HI, Yamada Y, Tikhonova EB, Ge Q, Krishnamoorthy G: Structural and functional diversity of bacterial membrane fusion proteins. *Biochim Biophys Acta* 2009, 1794:794-807.

[102] Bogomolnaya LM, Andrews KD, Talamantes M, Maple A, Ragoza Y, Vazquez-Torres A, Andrews-Polymenis H: The ABC-type efflux pump MacAB protects Salmonella enterica serovar typhimurium from oxidative stress. *MBio* 2013, 4:e00630-00613.

[103] Kuroda T, Tsuchiya T: Multidrug efflux transporters in the MATE family. *Biochim Biophys Acta* 2009, 1794:763-768.

[104] Bay DC, Rommens KL, Turner RJ: Small multidrug resistance proteins: a multidrug transporter family that continues to grow. *Biochimica et Biophysica Acta (BBA)-Biomembranes* 2008, 1778:1814-1838.

[105] Borges-Walmsley MI, Walmsley AR: The structure and function of drug pumps. *Trends in microbiology* 2001, 9:71-79.

[106] Grkovic S, Brown MH, Skurray RA: Transcriptional regulation of multidrug efflux pumps in bacteria. In *Seminars in cell & developmental biology*. Elsevier; 2001: 225-237.

[107] Kumar A, Schweizer HP: Bacterial resistance to antibiotics: active efflux and reduced uptake. *Advanced drug delivery reviews* 2005, 57:1486-1513.

[108] Paulsen IT, Skurray RA, Tam R, Saier MH, Turner RJ, Weiner JH, Goldberg EB, Grinius LL: The SMR family: a novel family of multidrug efflux proteins involved with the efflux of lipophilic drugs. *Molecular microbiology* 1996, 19:1167-1175.

[109] He GX, Landry M, Chen H, Thorpe C, Walsh D, Varela MF, Pan H: Detection of benzalkonium chloride resistance in community environmental isolates of staphylococci. *J Med Microbiol* 2014, 63:735-741.

[110] Van Bambeke F, Balzi E, Tulkens PM: Antibiotic efflux pumps. *Biochemical pharmacology* 2000, 60:457-470.
[111] Jack DL, Storms ML, Tchieu JH, Paulsen IT, Saier MH: A broad-specificity multidrug efflux pump requiring a pair of homologous SMR-type proteins. *Journal of bacteriology* 2000, 182:2311-2313.
[112] Zhang Z, Ma C, Pornillos O, Xiu X, Chang G, Saier MH: Functional characterization of the heterooligomeric EbrAB multidrug efflux transporter of Bacillus subtilis. *Biochemistry* 2007, 46:5218-5225.
[113] Putman M, van Veen HW, Konings WN: Molecular properties of bacterial multidrug transporters. *Microbiology and Molecular Biology Reviews* 2000, 64:672-693.
[114] Nikaido H: Multidrug efflux pumps of gram-negative bacteria. *Journal of Bacteriology* 1996, 178:5853.
[115] Prosecka J, Orlov AV, Fantin YS, Zinchenko VV, Babykin MM, Tichy M: A novel ATP-binding cassette transporter is responsible for resistance to viologen herbicides in the cyanobacterium Synechocystis sp. PCC 6803. *FEBS journal* 2009, 276:4001-4011.
[116] Fernández L, Hancock RE: Adaptive and mutational resistance: role of porins and efflux pumps in drug resistance. *Clinical microbiology reviews* 2012, 25:661-681.
[117] Fuentes DE, Navarro CA, Tantaleán JC, Araya MA, Saavedra CP, Pérez JM, Calderón IL, Youderian PA, Mora GC, Vásquez CC: The product of the qacC gene of Staphylococcus epidermidis CH mediates resistance to β-lactam antibiotics in Gram-positive and Gram-negative bacteria. *Research in microbiology* 2005, 156:472-477.
[118] Marquez B: Bacterial efflux systems and efflux pumps inhibitors. *Biochimie* 2005, 87:1137-1147.
[119] McKeegan KS, Borges-Walmsley MI, Walmsley AR: The structure and function of drug pumps: an update. *Trends in microbiology* 2003, 11:21-29.
[120] Saier Jr MH, Paulsen IT: Phylogeny of multidrug transporters. In *Seminars in cell & developmental biology*. Elsevier; 2001: 205-213.
[121] Saier MH: A functional-phylogenetic classification system for transmembrane solute transporters. *Microbiology and Molecular Biology Reviews* 2000, 64:354-411.
[122] Saier MH, Paulsen IT, Sliwinski MK, Pao SS, Skurray RA, Nikaido H: Evolutionary origins of multidrug and drug-specific efflux pumps in bacteria. *The FASEB Journal* 1998, 12:265-274.
[123] Paulsen I, Littlejohn T, Rådström P, Sundström L, Sköld O, Swedberg G, Skurray R: The 3'conserved segment of integrons contains a gene associated with multidrug resistance to antiseptics and disinfectants. *Antimicrobial agents and chemotherapy* 1993, 37:761-768.
[124] Srinivasan VB, Rajamohan G: KpnEF, a new member of the Klebsiella pneumoniae cell envelope stress response regulon, is an SMR-type efflux pump involved in broad-spectrum antimicrobial resistance. *Antimicrobial agents and chemotherapy* 2013, 57:4449-4462.
[125] Kücken D, Feucht H-H, Kaulfers P-M: Association of qacE and qacEΔ1 with multiple resistance to antibiotics and antiseptics in clinical isolates of Gram-negative bacteria. *FEMS Microbiology Letters* 2000, 183:95-98.
[126] Tosini F, Visca P, Luzzi I, Dionisi AM, Pezzella C, Petrucca A, Carattoli A: Class 1 integron-borne multiple-antibiotic resistance carried by IncFI and IncL/M plasmids in

Salmonella enterica serotype Typhimurium. *Antimicrobial agents and chemotherapy* 1998, 42:3053-3058.

[127] Kazama H, Hamashima H, Sasatsu M, Arai T: Distribution of the antiseptic-resistance genes qacE and qacEΔ1 in Gram-negative bacteria. *FEMS microbiology letters* 1998, 159:173-178.

[128] He GX, Zhang C, Crow RR, Thorpe C, Chen H, Kumar S, Tsuchiya T, Varela MF: SugE, a new member of the SMR family of transporters, contributes to antimicrobial resistance in Enterobacter cloacae. *Antimicrob Agents Chemother* 2011, 55:3954-3957.

[129] Rapp M, Granseth E, Seppälä S, Von Heijne G: Identification and evolution of dual-topology membrane proteins. *Nature structural & molecular biology* 2006, 13:112-116.

[130] Greener T, Govezensky D, Zamir A: A novel multicopy suppressor of a groEL mutation includes two nested open reading frames transcribed from different promoters. *The EMBO journal* 1993, 12:889.

[131] Sikora CW, Turner RJ: SMR proteins SugE and EmrE bind ligand with similar affinity and stoichiometry. *Biochemical and biophysical research communications* 2005, 335:105-111.

[132] Son MS, Del Castilho C, Duncalf KA, Carney D, Weiner JH, Turner RJ: Mutagenesis of SugE, a small multidrug resistance protein. *Biochemical and biophysical research communications* 2003, 312:914-921.

[133] Marger MD, Saier MH, Jr.: A major superfamily of transmembrane facilitators that catalyse uniport, symport and antiport. *Trends Biochem Sci* 1993, 18:13-20.

[134] Griffith JK, Baker ME, Rouch DA, Page MG, Skurray RA, Paulsen IT, Chater KF, Baldwin SA, Henderson PJ: Membrane transport proteins: implications of sequence comparisons. *Curr Opin Cell Biol* 1992, 4:684-695.

[135] Saier MH, Jr., Beatty JT, Goffeau A, Harley KT, Heijne WH, Huang SC, Jack DL, Jahn PS, Lew K, Liu J, et al.: The major facilitator superfamily. *J Mol Microbiol Biotechnol* 1999, 1:257-279.

[136] Barrett MP, Walmsley AR, Gould GW: Structure and function of facilitative sugar transporters. *Curr Opin Cell Biol* 1999, 11:496-502.

[137] Paulsen IT, Brown MH, Skurray RA: Proton-dependent multidrug efflux systems. *Microbiol Rev* 1996, 60:575-608.

[138] Kumar S, Lindquist IE, Sundararajan A, Rajanna C, Floyd JT, Smith KP, Andersen JL, He G, Ayers RM, Johnson JA, et al.: Genome Sequence of Non-O1 Vibrio cholerae PS15. *Genome Announc* 2013, 1.

[139] Mukherjee M, Kakarla P, Kumar S, Gonzalez E, Floyd JT, Inupakutika M, Devireddy AR, Tirrell SR, Bruns M, He G, et al.: Comparative genome analysis of non-toxigenic non-O1 versus toxigenic O1 Vibrio cholerae. *Genomics Discovery* 2014, 2.

[140] Luo Y, Kong Q, Yang J, Golden G, Wanda SY, Jensen RV, Ernst PB, Curtiss R, 3rd: Complete genome sequence of the universal killer Salmonella enterica Serovar Typhimurium UK-1 (ATCC 68169). *J Bacteriol* 2011, 193:4035-4036.

[141] McClelland M, Sanderson KE, Spieth J, Clifton SW, Latreille P, Courtney L, Porwollik S, Ali J, Dante M, Du F, et al.: Complete genome sequence of Salmonella enterica serovar Typhimurium LT2. *Nature* 2001, 413:852-856.

[142] Parkhill J, Dougan G, James KD, Thomson NR, Pickard D, Wain J, Churcher C, Mungall KL, Bentley SD, Holden MT, et al.: Complete genome sequence of a multiple drug resistant Salmonella enterica serovar Typhi CT18. *Nature* 2001, 413:848-852.

[143] Boyd D, Peters GA, Cloeckaert A, Boumedine KS, Chaslus-Dancla E, Imberechts H, Mulvey MR: Complete nucleotide sequence of a 43-kilobase genomic island associated with the multidrug resistance region of Salmonella enterica serovar Typhimurium DT104 and its identification in phage type DT120 and serovar Agona. *J Bacteriol* 2001, 183:5725-5732.

[144] Michael GB, Butaye P, Cloeckaert A, Schwarz S: Genes and mutations conferring antimicrobial resistance in Salmonella: an update. *Microbes Infect* 2006, 8:1898-1914.

[145] Henderson PJ: Proton-linked sugar transport systems in bacteria. *J Bioenerg Biomembr* 1990, 22:525-569.

[146] Henderson PJ: The homologous glucose transport proteins of prokaryotes and eukaryotes. *Res Microbiol* 1990, 141:316-328.

[147] Henderson PJ, Maiden MC: Homologous sugar transport proteins in Escherichia coli and their relatives in both prokaryotes and eukaryotes. *Philos Trans R Soc Lond B Biol Sci* 1990, 326:391-410.

[148] Floyd JL, Smith KP, Kumar SH, Floyd JT, Varela MF: LmrS is a multidrug efflux pump of the major facilitator superfamily from Staphylococcus aureus. *Antimicrob Agents Chemother* 2010, 54:5406-5412.

[149] Smith KP, Kumar S, Varela MF: Identification, cloning, and functional characterization of EmrD-3, a putative multidrug efflux pump of the major facilitator superfamily from Vibrio cholerae O395. *Arch Microbiol* 2009, 191:903-911.

[150] Varela MF, Griffith JK: Nucleotide and deduced protein sequences of the class D tetracycline resistance determinant: relationship to other antimicrobial transport proteins. *Antimicrob Agents Chemother* 1993, 37:1253-1258.

[151] Henderson PJ, Roberts PE, Martin GE, Seamon KB, Walmsley AR, Rutherford NG, Varela MF, Griffith JK: Homologous sugar-transport proteins in microbes and man. *Biochem Soc Trans* 1993, 21:1002-1006.

[152] Balleste-Delpierre C, Sole M, Domenech O, Borrell J, Vila J, Fabrega A: Molecular study of quinolone resistance mechanisms and clonal relationship of Salmonella enterica clinical isolates. *Int J Antimicrob Agents* 2014, 43:121-125.

[153] Chen S, Cui S, McDermott PF, Zhao S, White DG, Paulsen I, Meng J: Contribution of target gene mutations and efflux to decreased susceptibility of Salmonella enterica serovar typhimurium to fluoroquinolones and other antimicrobials. *Antimicrob Agents Chemother* 2007, 51:535-542.

[154] Fluman N, Bibi E: Bacterial multidrug transport through the lens of the major facilitator superfamily. *Biochim Biophys Acta* 2009, 1794:738-747.

[155] Hur J, Jawale C, Lee JH: Antimicrobial resistance of Salmonella isolated from food animals: A review. *Food Research International* 2012, 45:819-830.

[156] Abraham S, Groves MD, Trott DJ, Chapman TA, Turner B, Hornitzky M, Jordan D: Salmonella enterica isolated from infections in Australian livestock remain susceptible to critical antimicrobials. *Int J Antimicrob Agents* 2014, 43:126-130.

[157] Randall LP, Cooles SW, Osborn MK, Piddock LJ, Woodward MJ: Antibiotic resistance genes, integrons and multiple antibiotic resistance in thirty-five serotypes of

Salmonella enterica isolated from humans and animals in the UK. *J Antimicrob Chemother* 2004, 53:208-216.

[158] Dionisi AM, Lucarelli C, Benedetti I, Owczarek S, Luzzi I: Molecular characterisation of multidrug-resistant Salmonella enterica serotype Infantis from humans, animals and the environment in Italy. *Int J Antimicrob Agents* 2011, 38:384-389.

[159] Riano I, Moreno MA, Teshager T, Saenz Y, Dominguez L, Torres C: Detection and characterization of extended-spectrum beta-lactamases in Salmonella enterica strains of healthy food animals in Spain. *J Antimicrob Chemother* 2006, 58:844-847.

[160] Wong MH, Yan M, Chan EW, Biao K, Chen S: Emergence of Clinical Salmonella enterica Serovar Typhimurium Isolates with Concurrent Resistance to Ciprofloxacin, Ceftriaxone, and Azithromycin. *Antimicrob Agents Chemother* 2014, 58:3752-3756.

[161] Giraud E, Cloeckaert A, Kerboeuf D, Chaslus-Dancla E: Evidence for active efflux as the primary mechanism of resistance to ciprofloxacin in Salmonella enterica serovar typhimurium. *Antimicrob Agents Chemother* 2000, 44:1223-1228.

[162] Baucheron S, Le Hello S, Doublet B, Giraud E, Weill FX, Cloeckaert A: ramR mutations affecting fluoroquinolone susceptibility in epidemic multidrug-resistant Salmonella enterica serovar Kentucky ST198. *Front Microbiol* 2013, 4:213.

[163] Nishino K, Nikaido E, Yamaguchi A: Regulation of multidrug efflux systems involved in multidrug and metal resistance of Salmonella enterica serovar Typhimurium. *J Bacteriol* 2007, 189:9066-9075.

[164] Davin R A J-MB, Chloë E. James, Jean-Philippe Lavigne, Jacqueline Chevalier,, Eric Garnotel AMaJ-MP: Membrane permeability and regulation of drug "influx and efflux" in Enterobacterial pathogens. *Current Drug Targets* 2008, 9:750-759.

[165] Piddock LJV, White DG, Gensberg K, Pumbwe L, Griggs DJ: Evidence for an Efflux Pump Mediating Multiple Antibiotic Resistance in Salmonella enterica Serovar Typhimurium. *Antimicrob Agents Chemother* 2000, 44:3118-3121.

[166] Alekshun MN, Levy SB: The mar regulon: multiple resistance to antibiotics and other toxic chemicals. *Trends Microbiol* 1999, 7:410-413.

[167] Rickard AH, Lindsay S, Lockwood GB, Gilbert P: Induction of the mar operon by miscellaneous groceries. *J Appl Microbiol* 2004, 97:1063-1068.

[168] Peng Y, Hernandez RL, Crow RR, Jones SE, Mathews SA, Arnold AM, Castillo EF, Moseley JM, Varela MF: Pasteurized whole milk confers reduced susceptibilities to the antimicrobial agents trimethoprim, gatifloxacin, cefotaxime and tetracycline via the marRAB locus in Escherichia coli. *J Dairy Res* 2008, 75:491-496.

[169] Bailey AM, Ivens A, Kingsley R, Cottell JL, Wain J, Piddock LJ: RamA, a member of the AraC/XylS family, influences both virulence and efflux in Salmonella enterica serovar Typhimurium. *J Bacteriol* 2010, 192:1607-1616.

[170] Charani E, Cooke J, Holmes A: Antibiotic stewardship programmes--what's missing? *J Antimicrob Chemother* 2010, 65:2275-2277.

[171] Levy SB, Marshall B: Antibacterial resistance worldwide: causes, challenges and responses. *Nat Med* 2004, 10:S122-129.

[172] Levy SB: Antibiotic resistance: consequences of inaction. *Clin Infect Dis* 2001, 33 Suppl 3:S124-129.

[173] Owens RC, Jr.: Antimicrobial stewardship: concepts and strategies in the 21st century. *Diagn Microbiol Infect Dis* 2008, 61:110-128.

[174] Zechini B, Versace I: Inhibitors of multidrug resistant efflux systems in bacteria. *Recent Pat Antiinfect Drug Discov* 2009, 4:37-50.

[175] Lomovskaya O, Bostian KA: Practical applications and feasibility of efflux pump inhibitors in the clinic--a vision for applied use. *Biochem Pharmacol* 2006, 71:910-918.

[176] Kumar S, Mukherjee MM, Varela MF: Modulation of Bacterial Multidrug Resistance Efflux Pumps of the Major Facilitator Superfamily. *International Journal of Bacteriology* 2013, 2013:15.

[177] Neupane GP, Kim DM, Kim SH, Lee BK: In vitro synergism of ciprofloxacin and cefotaxime against nalidixic acid-resistant Salmonella enterica serotypes Paratyphi A and Paratyphi B. *Antimicrob Agents Chemother* 2010, 54:3696-3701.

[178] Tiwari R, Dhama K, Kumar A, Rahal A, Kapoor S: Bacteriophage therapy for safeguarding animal and human health: a review. *Pak J Biol Sci* 2014, 17:301-315.

In: Salmonella
Editor: Christopher B. Hackett

ISBN: 978-1-63463-651-3

Chapter 2

SALMONELLA HEIDELBERG: A MULTI-DRUG RESISTANT FOOD-BORNE PATHOGEN

Indu Upadhyaya, Samantha Fancher, Abhinav Upadhyay and Kumar Venkitanarayanan*

Department of Animal Science, University of Connecticut, Storrs, CT, US

ABSTRACT

Salmonella enterica serovar Heidelberg is one of the most commonly detected serovars in swine and poultry, ranking among the top five serotypes associated with human salmonellosis. Although *S.* Heidelberg outbreaks have been linked with several foods of animal origin, poultry-derived foods serve as the major source of human infections in the United States and Canada. Moreover, recent research has shown that this serotype of *Salmonella* is significantly more invasive in humans when compared to other non-typhoidal counterparts. Severe infections caused by *S.* Heidelberg require antibiotic therapy in both adults and children. However, *S.* Heidelberg strains resistant to multiple antibiotics, including streptomycin, kanamycin, ceftriaxone, ciprofloxacin and other beta-lactam antibiotics, have been isolated from foods, especially poultry meat, thereby raising concerns on poultry products serving as a reservoir of multidrug resistant food-borne salmonellosis in humans. This chapter discusses the emergence of *S.* Heidelberg as a significant foodborne pathogen with emphasis on its prevalence, virulence, antibiotic resistance, and public health significance. Additionally, potential strategies for controlling *S.* Heidelberg in poultry are highlighted.

* Corresponding author, 3636 Horsebarn Hill Rd Ext., Unit 4040, Storrs, CT, USA. Phone:(860)-486-0947; Fax:(860)-486-4375, Email:kumar.venkitanarayanan@uconn.edu.

1. INTRODUCTION

Foodborne infections due to non-typhoidal Salmonella (NTS) are a global public health concern. Annually, NTS-associated foodborne gastroenteritis results in an estimated 94 million illnesses and 115,000 deaths worldwide (CDC, 2013a). In the United States, despite a significant progress in improving food safety, approximately 1 million cases of NTS diarrhea, 19,000 hospitalizations, and 400 deaths occur each year (Scallan et al., 2011), costing the US economy approximately $365 million in healthcare (USDA-ERS, 2010). *Salmonella enterica* subsp. *enterica* serovar Heidelberg (*S.* Heidelberg) is one of the top 5 serovars causing human non-typhoidal salmonellosis in the U.S (CDC, 2011a; Han et al., 2011). A recent estimate suggests that *S.* Heidelberg causes approximately 84,000 annual illnesses in the US (Foley et al., 2011; Han et al., 2011), characterized by gastroenteritis, fever, and abdominal cramps that develop 12 to 72 hours after ingesting contaminated food. *S.* Heidelberg is also associated with invasive infections causing septicemia and myocarditis in humans (Wilmshurst and Sutcliffe, 1995). Presently, *S.* Heidelberg is the second most common cause of death due to NTS infections in the United States, after S. *enterica* serovar Typhimurium (Kennedy et al., 2004; Patchanee et al., 2008; Crump et al., 2011).

S. Heidelberg has a broad host range, including poultry and farm animals. The pathogen is frequently isolated from chickens, turkeys, and swine (Foley and Lynne, 2008; Zhao et al., 2008) with chicken meat, eggs, pork and ground turkey being the primary source of infections to humans (Chittick et al., 2006; NARMS 2010). In 2013, *S.* Heidelberg was implicated in several outbreaks and sporadic cases of gastroenteritis in the U.S. and Canada due to the consumption of contaminated chicken (CDC, 2014a). As many as 770 persons were reported to be infected with *S.* Heidelberg and approximately 35% of these cases were hospitalized (CDC, 2014a). Follow up laboratory investigations revealed that the outbreak strains of *S.* Heidelberg were resistant to several commonly prescribed antibiotics in humans (CDC, 2014a). Subsequent genome analysis of *S.* Heidelberg showed that the pathogen possesses several genes critical for colonization in food animals, survival in food processing environments, and causing infection in humans (Hoffman et al., 2014).

This chapter discusses the emergence of *S.* Heidelberg as a foodborne pathogen with an emphasis on its prevalence, virulence, antibiotic resistance, and public health significance. Additionally, potential strategies for the diagnosis and control of *S.* Heidelberg infections in poultry are highlighted.

2. PREVALENCE, OUTBREAKS AND SOURCE OF *S.* HEIDELBERG

2.1. Prevalence of *S.* Heidelberg

Since 1962, pathogenic *Salmonella* species have been under surveillance in the United States by the Council of State and Territorial Epidemiologists, the Association of Public Health Laboratories, and the Centers for Disease Control and Prevention (CDC, 2011a). Among the various Salmonellae, *S.* Heidelberg has emerged as one of the common serovars associated with foodborne infections in humans (Folster et al., 2012; Patchanee et al., 2008). Although *S.* Heidelberg infections account for only 3.5% of all reported *Salmonella* illnesses in the U.S., the pathogen is responsible for approximately 7% of all *Salmonella*-related deaths, which is the second highest after *S.* Typhimurium (Kennedy et al., 2004, CDC, 2011a). Moreover, the diagnostic frequency for this serovar in North America has been relatively higher than other *Salmonella* (Gebreyes et al., 2000; Currie et al., 2005; Liljebjelke et al., 2005, Chittick et al., 2006; Nayak et al., 2007; Foley et al., 2008). Another critical feature of *S.* Heidelberg is that it is associated with more invasive infections than those caused by other serotypes. Therefore, it is mandated to continuously monitor *S.* Heidelberg prevalence in the food supply, in addition to other *Salmonella* serotypes of public health importance (Hur et al., 2012). According to the World Health Organization's surveillance of global *Salmonella* distribution, *S.* Heidelberg was not among the most commonly isolated *Salmonella* serovars of human and animal origin in Europe, Asia, and Australia earlier in the decade (Galanis et al., 2006). However, since 2006 the incidence of *S.* Heidelberg has increased substantially.

2.2. Outbreaks

S. Heidelberg outbreaks have been traced back as early as 1973. Chittick and co-workers (2006) analyzed national foodborne outbreak data from 1973 through 2001 to determine the proportion of *S.* Heidelberg outbreaks caused by specific foods. It was reported that among 6,633 outbreaks with known etiology, 184 (3%) were attributed to *S.* Heidelberg. These researchers also confirmed that at least 53 outbreaks out of the 101 identified were poultry or egg-related, thereby underscoring the importance of interventions for reducing *S.* Heidelberg contamination in poultry and eggs. Another *S.* Heidelberg outbreak in New Mexico reported diarrheal illness among affected patients costing over $11,000 on medical costs in addition to loss of labor (Anonymous, 1986).

In the past few years, there have been numerous multistate, food-borne outbreaks of *S.* Heidelberg in the United States. In 2011, the CDC reported a multistate outbreak of *S.* Heidelberg linked to the consumption of contaminated ground turkey and broiled chicken livers that sickened 130 people in 34 states (CDC, 2011b). In 2013, a multistate outbreak of *S.* Heidelberg infection was linked to contaminated chicken (CDC, 2013b). Another outbreak was linked to the ingestion of mechanically separated chicken at a correctional facility in 2014 (CDC 2014a). The *S.* Heidelberg strains associated with the 2011 and several other outbreaks were found to be resistant to multiple antimicrobial agents, including ampicillin, tetracycline, and in some cases streptomycin and gentamicin (Hoffman et al., 2013). The most recent outbreak of multidrug-resistant *S.* Heidelberg infections was linked to a commercial brand of chicken affecting 524 people from 25 states (CDC 2014a). In addition to these outbreaks, *S.* Heidelberg resistant to multiple antibiotics, including resistance to clinically important third generation cephalosporins, have been detected on food products, and food processing environments (Aarestrup et al., 2004, Han et al., 2011, Zhao et al., 2008, Hoffman et al., 2013, Gokulan et al., 2013).

2.3. Sources of *S.* Heidelberg

According to the USDA's National Veterinary Services Laboratory *S.* Heidelberg is one of the most commonly detected serovars from turkey, chicken, and swine establishments (Gokulan et al., 2013, CDC, 2011a). In addition, *S.* Heidelberg is one of the frequently isolated serotypes from retail meats, reported as a part of the National Antimicrobial Resistance Monitoring System (NARMS) (FDA, 2010). Besides meat, egg consumption has been found to be a risk factor for infection with *S.* Heidelberg. This pathogen has been isolated from chicken manure, surface of eggs, and has been shown to penetrate the eggshell (Hennessey et al., 2004). Moreover, it has been recovered from the ovaries and peritoneum of egg-laying chickens (Hennessey et al., 2004, Poppe et al., 1992). Consequently the CDC has classified *S.* Heidelberg as a significant cause of egg-associated human disease (Chittick et al., 2004; Hennessy et al., 2004). Other pathogenic *Salmonella* serotypes previously implicated in egg-borne outbreaks include Enteritidis and Typhimurium (Angulo et al., 1998, CDC, 2000, Hedberg et al., 1993).

2.4. Comparative genomics of *Salmonella* Heidelberg

In recent years, *S.* Heidelberg has emerged as one of the most frequently isolated *Salmonella* serovars from poultry and poultry products in the United States (Liljebjelke et al., 2005; Kabir, 2010). Several factors, especially acquisition/loss of gene via plasmids and phages influence the host range of pathogens. With advancements in sequencing technologies the ability to compare genomes of emerging and established pathogens has greatly enhanced our capacity to discern such genetic exchange and characterize the factors critical for their survival and pathogenicity inside the host. Hoffman and coworkers (2014) characterized the genetic diversity of 46 *S.* Heidelberg isolates, including 9 strains associated with 2011 outbreak, using whole-genome sequencing. The authors observed that the outbreak strains clustered together with a mean single nucleotide polymorphism (SNP) variation of 17 SNPs, indicating significant clonal relatedness. In addition, the isolates had an array of phages, including the P22, P4, and Gifsy-2 that carry the majority of virulence genes. Moreover, 13 fimbrial markers and antibiotic resistance plasmids of the (Inc) I1, IncA/C, and IncHI2 groups that are critical for colonization and survival of *Salmonella* in the host were detected. Twenty-one strains contained IncX plasmid that carries a type IV secretion system. The genes coding for *Salmonella* Pathogenicity Island 1 and 2 have also been detected in other *S.* Heidelberg isolates from poultry production environments (Zou et al., 2010). In another study by Lynne and coworkers (2009), a significant genetic heterogeneity was observed in the antimicrobial resistance genes of strains isolated from different food sources or regions. In addition, a large number of the MDR genes were carried on plasmids, indicating that *S.* Heidelberg genome potentially evolved by a high degree of horizontal gene transfer.

Although commonly isolated from chicken, *S.* Heidelberg does not seem to have a strong host restriction. Several researchers compared *S.* Heidelberg strains isolated from human patients with those recovered from food animal species, and found a significant overlap in the PFGE patterns, plasmid types, and antimicrobial resistance profiles (Han et al., 2011; Mazurek et al., 2004; Kaldhone et al., 2008), suggesting that *S.* Heidelberg has the ability to infect multiple host species (Foley et al., 2008). However, unlike other broad host range NTS serotypes, a higher percentage (11% as compared to 6% of *S.* Typhimurium and *S.* Enteritidis) of *S.* Heidelberg infections are found to be invasive in nature (Vugia et al., 2004). Interestingly, a number of *S.* Heidelberg strains isolated from poultry related sources were found to harbor IncFIB plasmids critical for extraintestinal survival, previously identified in avian-pathogenic *E. coli* (Han et al., 2012, Didelot et al., 2011; Johnson and Nolan, 2009). In many cases, the IncFIB plasmids also carry genes that encode resistance to multiple antimicrobials, including streptomycin, chloramphenicol, and sulfonamide. The presence of such plasmids potentially increases the probability of co-selection of virulence genes along with antimicrobial resistance genes in pathogens exposed to antimicrobials commonly used as growth promoters in animals. In addition, *S.* Heidelberg carry genes encoding disinfectant and heavy metal resistance, which may facilitate its survival in the food-processing environment, where disinfectants such as chlorine are commonly used (Johnson et al., 2010; Kaldhone et al., 2008). However, further research is required to comprehensively understand the underlying molecular mechanisms that control *S.* Heidelberg infection process in animals and humans.

3. Molecular Determinants of Virulence in *S.* Heidelberg

Based on current literature, most studies on the virulence mechanisms of *Salmonella* have been performed on *S.* Typhimurium. Although a majority of the virulence mechanisms in *S.* Heidelberg are similar to those identified in *S.* Typhimurium, T4SS as a virulence factor critical for invasiveness is present only in *S.* Heidelberg (Gokulan et al., 2013). Furthermore, invasive *S.* Heidelberg, estimated at approximately 11% of *S.* Heidelberg isolates, can invade the intestinal epithelium through a process called bacterial-mediated endocytosis, primarily regulated by the T3SS1 (Ohl & Miller, 2001; Ibarra & Steele-Mortimer, 2009; Francis et al., 1992; Vugia et al., 2004). A few *S.* Heidelberg isolates can also survive in immune cells, in which T3SS2 is intimately involved leading to serious complications of the disease such as septicemia and organ systems infections (Gokulan et al., 2013; Sheppard et al., 2003). *S.* Heidelberg contains at least four transmissible plasmids (IncA/C, IncI, IncHI2 and IncX) that carry genes associated with virulence and resistance to antibiotics, heavy metals, or other disinfectants in addition to the virulent plasmids identified in *S.* Typhimurium (Hoffman et al., 2014).

3.1. Role of Virulence Factors in *S.* Heidelberg Infection Process

Similar to other *Salmonella* serotypes including *S.* Typhimurium, *S.* Heidelberg relies on fimbriae for attachment of the bacteria to the epithelial host cells and on flagella to facilitate bacterial motility and host cell invasion (Ibarra & Steele-Mortimer, 2009; Humphries *et al.*, 2001; Schmitt *et al.*, 2001). The role of *spvB* and *spvC* in destabilizing the cytoskeleton and host cell cytotoxicity (Ibarra & Steele-Mortimer, 2009; Lesnick *et al.*, 2001; Tezcan-Merdol *et al.*, 2001; Kurita *et al.*, 2003; Browne *et al.*, 2008) and inactivation of mitogen-associated protein kinases (Ibarra & Steele-Mortimer, 2009; Li *et al.*, 2007; Mazurkiewicz *et al.*, 2008) respectively, has been well established in *Salmonella* serotypes. However, a recent study observed that *S.* Heidelberg lack the *spvC* gene (Diarra et al., 2014).

S. Heidelberg possesses VirB4/D4 plasmids, which are associated with T4SS that has not been previously described in *Salmonella.* However, similar plasmid sequences are identified in *Shigella boydii, Citrobacter rodentium,* and *Escherichia coli* (Gokulan et al., 2013). A recent study reported that many *S.* Heidelberg isolates possessed the IncX plasmid, which houses genes containing structural components of the VirB4/D4 T4SS (Hoffman, 2014). The T4SS machinery can be grouped into 3 main components: 1) The pilus-associated units, 2) The channel-forming core unit, and 3) the coupling components consisting of ATPases. Through gene expression analysis, it was determined that the pilus-associated units were highly expressed at the initial stage of infection (5 minutes post-infection). The expression of genes associated with the core unit and coupling components was significantly increased at 30 minutes post infection compared to that of the early stage, and this was greater at 60 minutes post infection (Gokulan et al., 2013). These results indicated that *S.* Heidelberg containing the VirB4/D4 plasmids had an increased ability to lyse host cells. In the host cells that survived, such as macrophages, *S.* Heidelberg was persistent for a longer period of time. T4SS was also associated with a down-regulation of TLR signaling, apoptosis, and chemokine and cytokine signaling, which correlated with innate immune response suppression in hosts. Taken together, these findings suggest that T4SS plays a critical role in the survival of *S.* Heidelberg under hostile host environment (Gokulan et al., 2013).

4. Antibiotic Resistance in *S.* Heidelberg

Antibiotic-resistant *Salmonella* have been implicated with greater morbidity and mortality than pan-susceptible strains (Varma et al., 2005). As previously mentioned, the major virulence and antibiotic resistance mechanisms in *S.* Heidelberg are similar to those described in *S.* Typhimurium. However, multi-drug resistance in *S.* Heidelberg has been reported to increase from 12 percent in 1996 to 26 percent in 2009 only in the past decade (Medalla et al., 2013). In lieu of this increase in multi-drug resistance among *S.* Heidelberg, Hoffman (2014) studied the factors responsible for antibiotic resistance specifically in this serotype, which are described later.

The NARMS estimated that 33.9 percent of *S.* Heidelberg isolates were resistant to three or more antibiotics and 65 percent of *S.* Heidelberg isolates from an outbreak in ground turkey in 2010 were resistant to three or more antibiotics (NARMS, 2010, Hoffman, 2014). The present antibiotics that *S.* Heidelberg is most commonly resistant to are ceftriaxone, streptomycin, tetracycline, sulfamethoxazole, chloramphenicol, and trimethoprim-sulfamethoxazole (Hoffman, 2014, NARMS, 2010). The type of incompatibility (inc) plasmid present in *S.* Heidelberg strains determines their antibiotic resistance profile. For example, *S.* Heidelberg strains containing the IncA/C plasmid have been associated with antibiotic resistance genes such as *aphA, aacC, dfrA1, dfrA12, strA, strB, tetA, aadA, aadB,* bla_{CMY2}, bla_{TEM}, *sul1, sul2, cmlA1,* and *floR.* Similarly *S.* Heidelberg strains possessing the IncHI2 contain genes *neoR, strA, strB, aph, sph, tetA, tetC, tetD,* and *ble. S.* Heidelberg isolates containing the IncFIB plasmid have shown to incorporate *aphA, strA, strB, sul2, tetA, tetC* and *tetD* genes. Likewise, aacC, *aadA,* and *sul1* have been found in *S.* Heidelberg containing the IncI1 plasmid (Hoffman, 2014; Han et al., 2012). The clustering of antibiotic resistance genes with virulence genes on plasmids implies the potential horizontal transfer of these plasmids in *S.* Heidelberg (Han et al., 2012).

4.1. Antibiotic Resistance Mechanisms

In addition to the previously mentioned molecular mechanisms, the process of antibiotic resistance in *S.* Heidelberg has been described to be similar to those in *S.* Typhimurium. In general *S.* Typhimurium resistance to antimicrobials is attributed to several mechanisms such as degradation or structural modification by enzymes, reduction of bacterial cell permeability, activation of efflux pumps, and modification of the targets of antimicrobials (Hur, Jawale & Lee, 2011; Sefton, 2002). The resistance to penicillins and cephalosporins is due to the production of β-lactamase, which degrades their chemical structure (Hur, Jawale & Lee, 2011;Bush, 2003; Foley & Lynne, 2008). The resistance to quinolones and fluoroquinolones is attributed to the altering of DNA targets, the genes encoding for gyrase and topoisomerase, in addition to efflux pumps (Hur, Jawale & Lee, 2011; Eaves et al., 2004; Griggs, Gensberg & Piddock, 1996;Piddock, 2002; Piddock, Ricci, McLaren & Griggs, 1998;Heisig, 1993). Resistance to tetracycline and chloramphenicol is due to efflux pumps (Hur, Jawale & Lee, 2011; Butaye, Cloeckaert & Schwarz, 2003;Chopra & Roberts, 2001). Finally, resistance to sulfonamides and trimethoprim is often associated with an alteration in the folic acid biosynthesis pathway, which is targeted by these antibiotics (Hur, Jawale & Lee, 2011;Huovinen et al., 1995).

5. Detection and Control of *S.* Heidelberg

The ubiquitous distribution of *S.* Heidelberg in the environment, along with its ability to survive in soil for extended period of time (Holley et al., 2006) potentially leads to reoccurring introduction of the pathogen in farms, food animals, and the food processing environment. Therefore, development of good manufacturing practices, rapid detection methods, and the use of appropriate interventions are important for controlling the pathogen.

5.1. Detection Methods

Rapid detection methods are integral to controlling *Salmonella*-associated foodborne illnesses. According to the US Food and Drug Administration, any rapid detection method that identifies a targeted pathogen must be confirmed by traditional culture based methods (FDA, 2001). Culture based methods used for the identification of *S.* Heidelberg take 4 to 7 days to confirm a positive sample. Apart from being time consuming and labor intensive, cultural methods often fail to detect viable but non-culturable (VBNC) bacterial cells that could be potentially infectious. Due to serious health risks associated with *S.* Heidelberg infections, there exists a critical need in developing rapid molecular methods for detecting this pathogen. In this regard, a DNA-DNA hybridization assay was developed by Fitts and coworkers (1983), for rapid detection of *Salmonella* spp. (including *S*. Heidelberg) on foods. In another approach (Hong et al., 2008), a rapid multiplex PCR targeting the O and H antigen was developed for screening *S*. Eneritidis, Hadar and Heidelberg serotypes.

The advent of microarrays and next generation sequencing technology has provided new possibilities for developing rapid methods for pathogen identification. In a recent study, McCarthy and co-workers (2009) performed *in silico* genome comparison of *Salmonella* serovars to identify *S*. Heidelberg specific regions in the *Salmonella* genome. These regions were used to develop a Heidelberg-specific, PCR-based assay for rapid identification of *S*. Heidelberg.

5.2. Control of *S.* Heidelberg

With growing concerns over the use of antibiotics as growth promoters and increase in antimicrobial resistance in *S*. Heidelberg there is a renewed interest in exploring the potential of alternative approaches to control colonization of *S*. Heidelberg in food animals and reduce the contamination of food products. These approaches include use of natural bio-control agents such as plant-derived antimicrobials, probiotic bacteria, and bacteriophages. In addition, various vaccination strategies are being investigated for reducing *Salmonella* spp., including *S*. Heidelberg in poultry flocks. The major scientific studies that investigated the efficacy of aforementioned approaches for controlling *S*. Heidelberg in food animals, high-risk foods and food processing environment are summarized below.

5.2.1. Plant Extracts

Since ancient times, plant extracts have been used as food preservatives and dietary supplements to prevent food spoilage and maintain human health, respectively. The antimicrobial activity of several plant-derived compounds has been documented (Burt, 2004; Holley and Patel, 2005; Newman and Cragg, 2012) and an array of active components has been identified (Dixon, 2001). A majority of these compounds are secondary metabolites and are produced as a result of reciprocal interactions between plants, microbes, and animals (Reichling, 2010). They protect plants from pathogens, inter-plant competition, abiotic stresses (Briskin, 2000), and facilitate their survival in the local environment (Harborne, 1993). In addition to their use in traditional medicine, plant extracts have been used in ethno-veterinary practices for improving animal health and production (Greathead, 2003; Newman et al., 2000). The major advantage of using plant extracts for controlling pathogens in animals is that they do not exert deleterious effects on the health of animals. Moreover, evidence exists that the plant extracts do not induce resistance in the target pathogens (Ali et al., 2005; Ohno et al., 2003). A majority of plant-derived antimicrobials are generally recognized as safe (GRAS) compounds with low mammalian cytotoxicity and quick biodegradability in soil and water (Isman, 2000), thus making them safe and environment-friendly chemicals.

Several studies in the past have investigated the efficacy of plant-derived antimicrobials for controlling *S*. Enteritidis in broilers (Kollanoor-Johny et al., 2009; Kollanoor-Johny et al., 2012a,b) and layer chickens (Ordonez et al., 2008). However, only a few investigations have tested the effect of plant compounds on *S*. Heidelberg colonization in poultry. Amerah and coworkers (2012) investigated the effect of essential oils, cinnamaldehyde and thymol on broiler performance and *S*. Heidelberg colonization. These researchers found that in-feed supplementation of essential oils improved the 42 day body weight gain, feed efficiency, and significantly reduced the incidence of horizontal transmission of *S.* Heidelberg in the flock by 77%. In another study by Alali and coworkers (2013), supplementation of an essential oil blend containing carvacrol, thymol, eucalyptol and lemon at 0.05% in drinking water significantly reduced the colonization of *S.* Heidelberg in the crop of broiler chickens, and improved weight gain and feed conversion ratio in birds.

The complex microbiome of the chicken ceca plays a critical role in nutrient assimilation, pathogen colonization and overall health of birds (Qu et al., 2008). Changes in feed often alter the composition and diversity of gut microbial populations (Ley et al., 2006; Duncan et al., 2007), which in turn affects host metabolism (Brown et al., 2012) and susceptibility to gastrointestinal bacterial infections (Ghosh et al., 2011). In a recent study, Danzeisen and coworkers (2011) observed that supplementation of anticoccidial and growth promoting drugs such as virginiamycin and tylosin modulated the chicken cecal microbiome and metagenome. Similar modulation of cecal microbiota was observed when broiler diet was supplemented with antimicrobial feed additives, avilamycin and flavophospholipol (Torok et al., 2011). The modulation of gut microbiota has also been observed with dietary supplementation of flavonoids in mice (Espley et al., 2014). These results indicate that the plant-derived antimicrobials, in addition to their direct antimicrobial action on pathogen, could potentially act through modulation of the gut microbiota to reduce pathogen colonization. However, further research is required to elucidate the mechanism of their action on gut microbiota and pathogen in the host.

Besides the pre-harvest application in birds, several studies have investigated the efficacy of a variety of plant-derived antimicrobials and extracts in reducing *Salmonella* on poultry products (Upadhyaya et al., 2013, Mani Lopez et al., 2012, Pathania et al., 2010). Although plant compounds have the potential to control *S.* Heidelberg post harvest on poultry products, in-depth studies on their efficacy and impact on the sensory characteristics of meat and eggs are warranted.

5.2.2. Probiotic Bacteria

The Food and Agriculture Organization (FAO) and World Health Organization (WHO) define probiotics as "live microorganisms which, when administered in adequate amounts, confer a health benefit on the host". Probiotic bacteria facilitate nutrient assimilation (Sonnenburg et al., 2006; Yatsunenko et al., 2012), potentiate host immune function (Olszak et al., 2012), and protect against enteric pathogens (Candela et al., 2008; Fukuda et al., 2011). This is mediated by physical interactions, chemical cross talk, and the production of inhibitory metabolites that modulate inter-species bacterial colonization and survival in a niche or host (Carpentier and Chassaing, 2004; Kives et al., 2005; Valle et al., 2006). Several studies have investigated the efficacy of various lactic acid bacteria for controlling *S.* Enteritidis (Vicente et al., 2008; Tellez et al., 2001; Higgins et al., 2007), S. Typhimurium (Promsopone et al., 1998; Waters et al., 2005; Revolledo et al., 2009) and *Campylobacter jejuni* (Morishita et al., 1997) in poultry. However, only limited studies have investigated the effect of probiotics on *S.* Heidelberg colonization in poultry. In a recent investigation, Menconi and coworkers (2011) found that a commercial Lactobacillus-based probiotic culture mixture (FloraMax, IVS-Wynco LLC, Springdale, AR) significantly reduced *S.* Heidelberg colonization in cecal tonsils of broiler chicks and turkey poults. In another study, direct fed microbials (Primalac) containing *Lactobacillus acidophilus* and *L. casei* strains were found to exert an inhibitory effect on the intestinal colonization of turkey poult with *S.* Heidelberg (Grimes et al., 2008).

Apart from their role as a natural bio-control agent in susceptible hosts, probiotic bacteria and/or their antimicrobial metabolites are a viable and safe (GRAS status) option for inhibiting pathogenic microorganisms on food products (Holzapfel et al., 1995; Ross et al., 2002). The various metabolites produced by these bacteria such as bacteriocins, hydrogen peroxide and organic acids are used for preserving food products and extending shelf-life (Galvez et al., 2007). Kim and coworkers (2014) isolated bacteriocin producing lactic acid bacteria from broilers and found that they exerted a significant antimicrobial effect against major pathogenic *Salmonella* serovars, including *S.* Heidelberg.

5.2.3. Bacteriophages

Lytic bacteriophages are known to selectively infect and kill bacteria (Skurnik and Strauch, 2006) and are a potential tool for controlling foodborne pathogens such as *Salmonella* and *Campylobacter* (Goode et al., 2003; Higgins et al., 2005; Modi et al., 2001; Leverentz et al., 2001). Since bacteriophages are host specific other microbial populations are not affected. In a recent study Borie and coworkers (2008) isolated bacteriophages from the sewage system of commercial chicken flocks. These isolates were found to possess bactericidal activity against *S.* Heidelberg, *S.* Enteritidis and *S.* Agona. The application of the isolated bacteriophages as an aerosol spray or in drinking water reduced *S.* Enteritidis colonization in 10-day-old white leghorn chicks. In another study, bacteriophages P22 and 29C that were lytic against *S.* Enteritidis were found to exhibit significant attachment to *S.* Heidelberg isolates and reduced their population on inoculated chicken pieces (Goode et al., 2003). The authors concluded that the phages that attach to a common lipopolysaccharide antigen, such as O-12, might therefore be used effectively to lyse a majority of *Salmonella* strains within Kauffman-White groups A, B, and D. Similar findings were reported by Bardina and group (2012) who observed that bacteriophages most effective against *S.* Enteritidis and Typhimurium exhibited significant lytic efficacy against *S.* Virchow, *S.* Hadar and *S.* Infantis. Although there is possibility of emergence of phage-resistance in *Salmonella*, little is known about the frequency of this development. Therefore further research is critical before selecting phages for potential application in poultry or poultry products. Phage genome analysis is one such tool that facilitates identification of phages with the appropriate integration machinery that facilitates safe and effective application of phages as bio-control agents.

5.2.4. Vaccination

Unlike natural biocontrol agents, which rely on their interaction with pathogenic microbes for exerting antimicrobial effect, vaccination of chickens harnesses the bird's immune system to decrease the levels of target pathogen in the flock. Although promising, vaccination of poultry, especially broilers, presents a unique challenge due to the short life span of 5-6 weeks. A number of live attenuated, inactivated or subunit vaccines have been targeted against *S.* Enteritidis (Cooper et al., 1994; Hassan et al., 1994; Cerquetti et al., 2000; Methner et al., 2011; Nandre et al., 2012; Matulova et al., 2012) and cross protection to other serotypes has been documented. Deguchi and coworkers (2009) tested the efficacy of trivalent (*S.* Enteritidis, Typhimurium, Infantis) vaccine against shedding of *Salmonella* in chickens, and found that it provided cross immunity against *S.* Heidelberg and reduced fecal shedding of the pathogen. Similar results were obtained in another study, where vaccination with a live-avirulent *S.* Typhimurium strain provided cross protection against *S.* Enteritidis, *S.* Agona, *S.* Bredeney and *S.* Heidelberg in immunized chickens (Hassan et al., 1994).

Since the protection provided by inactivated or avirulent whole cell vaccines are usually of short duration, and may impart adverse side effects due to lipopolysaccharide components, the development of vaccines based on non-toxic and well-defined bacterial cellular components is underway. Charles and coworkers (1993) developed an anti-*S.* Heidelberg vaccine using outer membrane protein of the pathogen and incorporating it into lipid conjugated immunostaining complexes. The vaccine administration significantly reduced the isolation rate of *Salmonella* serotypes (*S.* Heidelberg, *S.* Reading, *S.* Enteritidis) from vaccinated turkeys as compared to controls. In spite of its advantages, the development of poultry vaccines face a number of challenges such as high cost of production, application and limited duration of protection.

Conclusion and Future Directions

Salmonella Heidelberg has emerged as a foodborne pathogen with significant public health and economic impact. The occurrence of multiple drug resistance in *S.* Heidelberg coupled with its high invasiveness and ability to produce severe extra-intestinal infections underscore the health impact of this pathogen in humans. Implementation of strict on-farm biosecurity measures for reducing *S.* Heidelberg colonization in food animals, judicious use of antibiotics in animal agriculture, and effective interventions for reducing post-harvest contamination of animal food products are critical for controlling food-borne transmission of this pathogen.

References

Aarestrup F. M., Hasman, H., Olsen, I., Sorensen, G. (2004). International spread of *bla* (CMY-2)-mediated cephalosporin resistance in a multiresistant *Salmonella enterica* serovar Heidelberg isolate stemming from the importation of a boar by Denmark from Canada. *Antimicrobial Agents Chemotherapy, 48,* 1916-1917.

Alali, W. Q., Hofacre, C. L., Mathis, G. F., & Faltys, G. (2013). Effect of essential oil compound on shedding and colonization of *Salmonella enterica* serovar Heidelberg in broilers. *Poultry Science, 92*, 836-841.

Ali, S. M., Khan, A. A., Ahmed, I., Musaddiq, M., Ahmed, K. S., Polasa, H., Rao, V.L., Habibullah, C.M., Sechi, L.A. & Ahmed, N. (2005). Antimicrobial activities of Eugenol and Cinnamaldehyde against the human gastric pathogen *Helicobacter pylori. Annals of Clinical Microbiology and Antimicrobials, 4,* 20.

Amerah, A. M., Mathis, G., & Hofacre, C. L. (2012). Effect of xylanase and a blend of essential oils on performance and *Salmonella* colonization of broiler chickens challenged with *Salmonella* Heidelberg. *Poultry Science, 91*, 943-947.

Angulo, F.J., Swerdlow, D.L. (1998). *Salmonella* Enteritidis infections in the United States [review]. *Journal of American Veterinary Medical Association*, 213, 1729–31.

Anonymous. (1986) Epidemiologic Notes and Reports *Salmonella* Heidelberg Outbreak at a Convention -- New Mexico. *MMWR Weekly Reports*, 35, 91.

Bardina, C., Spricigo, D. A., Cortes, P., & Llagostera, M. (2012). Significance of the bacteriophage treatment schedule in reducing *Salmonella* colonization of poultry. *Applied Environmental Microbiology, 78*, 6600-6607.

Borie, C., Albala, I., Sànchez, P., Sánchez, M. L., Ramírez, S., Navarro, C., Morales, M. A. (2008). Bacteriophage treatment reduces *Salmonella* colonization of infected chickens. *Avian Diseases, 52*, 64-67.

Briskin, D. (2000). Medicinal plants and phytomedicines. Linking plant biochemistry and physiology to human health. *Plant Physiology, 124*, 507-514.

Brown, R. K., Ilhan, Z., Kank, D., & DiBaise, J. K. (2012). Effects of gut microbes on nutrient absorption and energy regulation. *Nutrition in Clinical Practice, 27*, 201-214.

Browne, S.H., Hasegawa, P., Okamoto, S., Fierer, J., & Guiney, D.G. (2008). Identification of *Salmonella* SPI-2 secretion system components required for *SpvB*-mediated cytotoxicity in macrophages and virulence in mice. *FEMS Immunology and Medical Microbiology, 52*, 194–201.

Burt, S. (2004). Essential oils: their antibacterial properties and potential applications in food. *International Journal of Food Microbiology, 94*, 223-53.

Bush, K. (2003). Beta-lactam antibiotics: Penicillins. In Finch, R. G., Greenwood, D., Norrby, S. R., & Whitley, R. J. (Eds.), Antibiotic and chemotherapy: Anti-infective agents and their use in therapy. (224-258). Edinburgh, UK: Churchill Livingstone.

Butaye, P., Cloeckaert, A. & Schwarz, S. (2003). Mobile genes coding for efflux-mediated antimicrobial resistance in gram-positive and gram-negative bacteria. *International Journals of Antimicrobial Agents, 22*, 205-210.

Candela, M., Perna, F., Carnevali, P., Vitali, B., Ciati, R., Gionchetti, P., Rizello, F. (2008). Interaction of probiotic *Lactobacillus* and *Bifidobacterium* strains with human intestinal epithelial cells: Adhesion properties, competition against enteropathogens and modulation of IL-8 production. *International Journal of Food Microbiology, 125*, 286-292.

Carpentier, B. & Chassaing, D. (2004). Interactions in biofilms between *Listeria monocytogenes* and resident microorganisms from food industry premises. *International Journal of Food Microbiology, 97*, 111-122.

Centers for Disease Control and Prevention (CDC). (2000) Outbreaks of *Salmonella* serotype Enteritidis infection associated with eating raw or undercooked shell eggs—United States, 1996–1998. *Morbity and Mortality Weekly Reports, 49,* 73-79.

Centers for Disease Control and Prevention (CDC). (2011a). *Salmonella* Surveillance: Annual Summary, 2009. Atlanta, GA: Centers for Disease Control and Prevention.

Centers for Disease Control and Prevention (CDC). (2011b). Multistate Outbreak of Human *Salmonella* Heidelberg Infections Linked to Ground Turkey. Available at http://www.cdc.gov/salmonella/heidelberg/080111/index.html?s_cid=tw_cdc745. Accessed on 12 October 2014.

Centers for Disease Control and Prevention (CDC). (2013a). Antibiotic Resistance Threats in the United States. Available at http://www.cdc.gov/drugresistance/threat-report-2013/pdf/ar-threats-2013-508.pdf. Accessed on 12 October 2014.

Centers for Disease Control and Prevention (CDC). (2013a). Available at http://wwwnc.cdc.gov/travel/yellowbook/2014/chapter-3-infectious-diseases-related-to-travel/salmonellosis-nontyphoidal. Accessed 21 September, 2014.

Centers for Disease Control and Prevention (CDC). (2013b). Multistate Outbreak of Human *Salmonella* Heidelberg Infections Linked to Ground Turkey. Available at http://www.cdc.gov/salmonella/heidelberg-02-13/index.html. Accessed at 12 october 2014.

Centers for Disease Control and Prevention, CDC, 2014a. Multistate outbreak of *Salmonella* Heidelberg infections linked to chicken (Final Update). Available at http://www.cdc.gov/salmonella/heidelberg-02-13/index.html. Accessed on 12 october 2014.

Cerquetti, M. C. & Gherardi, M. M. (2000). Orally administered attenuated *Salmonella enteritidis* reduces chicken cecal carriage of virulent *Salmonella* challenge organisms. *Veterinary Microbiology, 76* , 185-192.

Chittick, P., Sulka, A., Tauxe, R. V., & Fry, A. M. (2004). Outbreaks of *Salmonella* Heidelberg infections in the United States: are eggs a common source. In *Proceedings of International Conference on Emerging Infectious Diseases,* Atlanta, GA, 179.

Chittick, P., Sulka, A., Tauxe, R. V., & Fry, A. M. (2006). A summary of national reports of foodborne outbreaks of *Salmonella* Heidelberg infections in the United States: clues for disease prevention. *Journal of Food Protection*, *69*, 1150-1153.

Chopra, I. & Roberts, M. (2001). Tetracycline antibiotics: Mode of action, applications, molecular biology, and epidemiology of bacterial resistance. *Microbiology and Molecular Biology Reviews, 65,* 232-260.

Cooper, G. L., Venables, L. M., Woodward, M. J., & Hormaeche, C. E. (1994). Vaccination of chickens with strain CVL30, a genetically defined *Salmonella* Enteritidis aroA live oral vaccine candidate. *Infection and Immunity*, *62*, 4747-4754.

Currie, A., MacDougall, L., Aramini, J., Gaulin, C., Ahmed, R., & Isaacs, S. (2005). Frozen chicken nuggets and strips and eggs are leading risk factors for *Salmonella* Heidelberg infections in Canada. *Epidemiology and infection, 133*, 809-816.

Danzeisen, J.L., Kim, H.B., Isaacson, R.E., Tu, Z.J., & Johnson, T.J. (2011). Modulations of the chicken cecal microbiome and metagenome in response to anticoccidial and growth promoter treatment. *PLoS ONE 6* (11), e27949. doi:10.1371/journal.pone.0027949.

Deguchi, K., Yokoyama, E., Honda, T., & Mizuno, K. (2009). Efficacy of a novel trivalent inactivated vaccine against the shedding of *Salmonella* in a chicken challenge model. *Avian Diseases, 53*, 281-286.

Diarra, M. S., Delaquis, P., Rempel, H., Bach, S., Harlton, C., Aslam, M., Pritchard, J. (2014). Antibiotic resistance and diversity of *Salmonella enterica* serovars associated with broiler chickens. *Journal of Food Protection, 77*, 40-9.

Didelot, X., Bowden, R., Street, T., Golubchik, T., Spencer, C., McVean, G., Sangal, V., Anjum, M.F., Achtman, M., Falush, D., & Donnelly, P. (2011). Recombination and population structure in *Salmonella enterica. PLoS Genetics, 7,* e1002191, doi:10.1371/journal.pgen.1002191.

Dixon, R. A. (2001). Natural products and plant disease resistance. *Nature, 411,* 843-847.

Duncan, S. H., Belenguer, A., Holtrop, G., Johnstone, A. M., Flint, H. J., & Lobley, G. E. (2007). Reduced dietary intake of carbohydrates by obese subjects results in decreased concentrations of butyrate and butyrate-producing bacteria in feces. *Applied and Environmental Microbiology, 73*, 1073-1078.

Eaves, D. J., Randall, L., Gray, D. T., Buckley, A., Woodward, M. J., White, A. P., & Piddock, L. J. V. (2004). Prevalence of mutations within the quinolone resistance-determining region of *gyrA, gyrB, parC*, and *parE* and association with antibiotic resistance in quinolone-resistant *Salmonella enterica. Antimicrobial Agents and Chemotherapy, 48,* 4012-4015.

Espley, R. V., Butts, C. A., Laing, W. A. Martell, S., Smith, H., McGhie, T. K., Zhang, J. (2014). Dietary flavonoids from modified apple reduce inflammation markers and modulate gut microbiota in mice. *Journal of Nutrition, 144*, 146-154.

Fitts, R., Diamond, M. Hamilton, C., & Nerri, M. (1983). DNA-DNA hybridization assay for detection of *Salmonella* spp. in foods. *Applied Environmental Microbiology, 46*, 1146-1151.

Foley, S. L., Lynne, A. M., & Nayak, R. (2008). *Salmonella* challenges: Prevalence in swine and poultry and potential pathogenicity of such isolates. *Journal of Animal Science, 86,* E149-162.

Folster, J. P., Pecic, G., Singh, A., Duval, B., Rickert, R., Ayers, S., & McDermott, P. F. (2012). Characterization of Extended-Spectrum Cephalosporin–Resistant *Salmonella enterica* serovar Heidelberg Isolated from Food Animals, Retail Meat, and Humans in the United States 2009. *Foodborne pathogens and disease, 9*, 638-645.

Food and Drug Administration (FDA). (2001). Archived BAM Method: Rapid Methods for Detecting Foodborne Pathogens. Available at http://www.fda.gov/Food/FoodScience Research/LaboratoryMethods/ucm109652.htm. Accessed 16 October, 2014.

Food and Drug Administration (FDA). (2010). National Antimicrobial Resistance Monitoring System-Enteric Bacteria (NARMS): 2007 Executive Report.

Francis, C. L., Starnbach, M. N., Falkow, S. (1992). Morphological and cytoskeletal changes in epithelial cells occur immediately upon interaction with *Salmonella typhimurium* grown under low-oxygen conditions. *Molecular Microbiology, 6*, 3077-3087.

Fukuda, S., Toh, H., Hase, K., Oshima, K., Nakanishi, Y., Yoshimura, K., Tobe, T. (2011). *Bifidobacteria* can protect from enteropathogenic infection through production of acetate. *Nature, 469,* 543-547.

Galanis, E., Wong, D. M. L. F., Patrick, M. E., Binsztein, N., Cieslik, A., Chalermchaikit, T., & Wegener, H. C. (2006). Web-based surveillance and global *Salmonella* distribution, 2000–2002. *Emerging infectious diseases, 12*, 381.

Galvez, A., Abriouel, H., Lopez, R. L., & Omar, N. B. (2007). Bacteriocin-based strategies for food biopreservation. *International Journal of Food Microbiology, 120*, 51-70.

Gebreyes, W. A., Davies, P. R., Morrow, W. M., Funk, J. A., & Altier, C. (2000). Antimicrobial Resistance of *Salmonella* Isolates from Swine. *Journal of clinical microbiology, 38*, 4633-4636.

Ghosh, S., Dai, C., Brown, K., Rajendiran, E., Makarenko, S., Baker, J. Ma, C. (2011). Colonic microbiota alters host susceptibility to infectious colitis by modulating inflammation, redox status, and ion transporter gene expression. *Gastrointestinal and Liver Physiology, 301*, G39-G49.

Gokulan K., Khare, S., Rooney, A. W., Han, J., Lynne, A. M., & Foley, S. L. (2013). Impact of plasmids, including those encoding VirB4/D4 Type IV Secretion Systems, on

Salmonella enterica serovar Heidelberg virulence in macrophages and epithelial cells. *PLoS ONE, 8*, e77866. doi:10.1371/journal.pone.0077866.

Goode, D., Allen, V. M., & Barrow, P. A. (2003). Reduction of experimental *Salmonella* and campylobacter contamination of chicken skin by application of lytic bacteriophages. *Applied and Environmental Microbiology, 69*, 5032-5036.

Greathead, H. (2003). Plants and plant extracts for improving animal productivity. *Proceedings of the Nutrition Society, 62,* 279-290.

Griggs, D. J., Gensberg, K. & Piddock, L. J. (1996). Mutations in *gyrA* gene of quinolone-resistant *Salmonella* serotypes isolated from humans and animals. *Antimicrobial Agents and Chemotherapy, 40,* 1009-1013.

Grimes J. L., Rahimi, S., Oviedo, E., Sheldon, B. W., & Santos, F. B. O. (2008). Effects of a direct-fed microbial (primalac) on turkey poult performance and susceptibility to oral *Salmonella* challenge. *Poultry Science, 87*, 1464-1470.

Han, J., David, D.E., Deck, J., Lynne, A.M., Kaldhone, P. (2011). Comparison of *Salmonella enterica* serovar Heidelberg isolates from human patients with those from animal and food sources. *Journal of Clinical Microbiology, 49,* 1130-1133.

Han, J., Lynne, A. M., David, D. E., Tang, H., Xu, J., Nayak, R., Kaldhone, P. (2012). DNA sequence analysis of plasmids from multidrug resistant *Salmonella enterica* serotype Heidelberg Isolates. *PLoS ONE 7*(12), e51160. doi:10.1371/journal.pone.0051160.

Harborne, J. B. (1993). Advances in chemical ecology. *Natural product reports, 10*, 327-348.

Hassan, J. O. & Curtiss, R. (1994). Development and evaluation of an experimental vaccination program using a live avirulent *Salmonella* typhimurium strain to protect immunized chickens against challenge with homologous and heterologous *Salmonella* serotypes. *Infection and Immunity, 62*, 5519-5527.

Hedberg, C.W., David, M.J., White, K.E., MacDonald, K.L., & Osterholm, M.T. (1993). Role of egg consumption in sporadic *Salmonella* Enteritidis and *Salmonella* Typhimurium infections in Minnesota. *Journal of Infectious Disease, 167*, 107–11.

Heisig, P. (1993). High-level fluoroquinolone resistance in a *Salmonella* Typhimurium isolate due to alterations in both *gyrA* and *gyrB* genes. *Journal of Antimicrobial Chemotherapy, 32,* 367-377.

Hennessy, T. W., Cheng, L. H., Kassenborg, H., Ahuja, S. D., Mohle-Boetani, J., Marcus, R., & Angulo, F. J. (2004). Egg consumption is the principal risk factor for sporadic *Salmonella* serotype Heidelberg infections: a case-control study in FoodNet sites. *Clinical Infectious Diseases, 38*(Supplement 3), S237-S243.

Herikstad, H., Vugia, D., Hadler, J. (1998). Population-based estimate of the burden of diarrheal illness: FoodNet 1996–1997 [abstract SS2]. (1998). Program and abstracts of the International Conference on Emerging Infectious Diseases.

Higgins, S. E., Erf, G. F., Higgins, J. P., Henderson, S. N., Wolfenden, A. D., Gaona-Ramirez, G., & Hargis, B. M. (2007). Effect of probiotic treatment in broiler chicks on intestinal macrophage numbers and phagocytosis of *Salmonella* Enteritidis by abdominal exudate cells. *Poultry Science, 86*, 2315-2321.

Higgins, S. E., Wolfendum, A. D., Bielke, L. R., Pixley, C. M., Torres-Rodriguez, A., Vicente, J. L., Bosseau, D. (2005). Application of ionized reactive oxygen species for disinfection of carcasses, table eggs, and fertile eggs. *The Journal of Applied Poultry Research, 14*, 716-720.

Hoffmann, M., Luo, Y., Lafon, P. C., Timme, R., Allard, M. W., McDermott, P. F., & Zhao, S. (2013). Genome sequences of *Salmonella enterica* serovar Heidelberg isolates isolated in the United States from a multistate outbreak of human Salmonella infections. *Genome announcements*, *1*(1), e00004-12.

Hoffmann, M., Zhao, S., Pettengill, J., Luo, Y., Monday, S. R., Abbott, J., ... & McDermott, P. F. (2014). Comparative Genomic Analysis and Virulence Differences in Closely Related Salmonella enterica Serotype Heidelberg Isolates from Humans, Retail Meats, and Animals. *Genome biology and evolution*, *6*, 1046-1068.

Holley, R. A., & Patel, D. (2005). Improvement in shelf-life and safety of perishable foods by plant essential oils and smoke antimicrobials. *Food Microbiology, 22*, 273-292.

Holley, R. A., Arrus, K. M., Ominski, K. H., Tenuta, M., & Blank, G. (2006). *Salmonella* survival in manure-treated soilds during simulated seasonal temperature exposure. *Journal of Environmental Quality, 35*, 1170-1180.

Holzapfel W. H., Geisen, R., & Schillinger, U. (1995). Biological preservation of foods with reference to protective cultures, bacteriocins and food-grade enzymes. *International Journal of Food Microbiology*, *24,* 343-362.

Hong, Y., Liu, T., Lee, M. D., Hofacre, C. L., Maier, M., David G White, D., M., & Ayers, S. (2008). Rapid screening of *Salmonella eenterica* serovars Enteritidis, Hadar, Heidelberg and Typhimurium using a serologically-correlative allelotyping PCR targeting the O and H antigen alleles. *BMC Microbiology, 8,* 178.

Humphries, A.D., Townsend, S.M., Kingsley, R.A., Nicholson, T.L., Tsolis, R.M., & Baumler, A.J. (2001). Role of fimbriae as antigens and intestinal colonization factors of *Salmonella* serovars. *FEMS Microbiology Letters, 201*, 121–125.

Huovinen, P., Sundstrom, L., Swedberg, G., & Skold, O. (1995). Trimethoprim and sulfonamide resistance. *Antimicrobial Agents and Chemotherapy, 39,* 279-289.

Hur, J., Jawale, C., & Hwa Lee, J. (2012). Antimicrobial resistance of *Salmonella* isolated from food animals: a review. *Food Research International, 45*, 819-830.

Ibarra, J. A. & Steele-Mortimer, O. (2009). *Salmonella* – the ultimate insider. *Salmonella* virulence factors that modulate intracellular survival. *Cellular Microbiology, 11*(11), 1579-1586.

Isman, M. B. (2000). Plant essential oils for pest and disease management. *Crop Protection, 19* (8-10), 603-608.

Johnson, T. J., & Nolan, L. K. (2009). Plasmid replicon typing. *Molecular Epidemiology of Microorganisms, 551,* 27-35.

Johnson, T.J., Thorsness, J.L., Anderson, C.P., Lynne, A.M., Foley, S.L., Han, J., Fricke, W.F., McDermott, P.F., White, D.G., Khatri, M., Stell, A.L., Flores, C., & Singer, R.S. (2010). Horizontal gene transfer of a ColV plasmid has resulted in a dominant avian clonal type of *Salmonella enterica* serovar Kentucky. *PLoS One 5*, e15524.

Kabir, S. M. (2010). Avian colibacillosis and salmonellosis: a closer look at epidemiology, pathogenesis, diagnosis, control and public health concerns. *International journal of environmental research and public health*, *7*, 89-114.

Kaldhone, P., Nayak, R., Lynne, A. M., David, D. E., McDermott, P. F. (2008) Characterization of *Salmonella enterica* serovar Heidelberg from Turkey-Associated Sources. *Applied Environmental Microbiology*, 74: 5038–46.

Kennedy, M., Villar, R., Vugia, D. J., Rabatsky-Her, T., Farley, M. M. (2004). Hospitalizations and deaths due to *Salmonella* infections, FoodNet, 1996-1999. *Clinical Infectious Disease 38 Suppl 3*: S142-S148.

Kim, J. Y., Young, J. A., Gunther, N. W., & Lee, J. (2014) Inhibition of *Salmonella* by bacteriocin-producing lactic acid bacteria derived from U.S. kimchi and broiler chicken. *Journal of Food Safety*, retrieved from http://onlinelibrary.wiley.com/doi/10.1111/jfs.12141/abstract, 10.1111/jfs.12141

Kives, J., Guadarrama, D., Orgaz, B., Rivera-Sen, A., Vazquez, J., & SanJose, C. (2005). Interactions in biofilms of *Lactococcus lactis* ssp. *cremoris* and *Pseudomonas fluorescens* cultured in cold UHT milk. *Journal of Dairy Science, 88*, 4165-4171.

Kollanoor-Johny, A., Baskaran, S. A., Charles, A. S., Amalaradjou, M. R., Darre, M. J., Khan, M. I., Hoagland, T. A. (2009). Prophylactic supplementation of caprylic acid in feed reduces *Salmonella* Enteritidis colonization in commercial broiler chicks. *Journal of Food Protection, 4,* 696-914.

Kollanoor-Johny, A., Mattson, T., Baskaran, S. A., Amalaradjou, M., Babapoor, S., March, B., Valipe, S. (2012a). Reduction of *Salmonella enterica* serovar Enteritidis colonization in 20-day-old broiler chickens by the plant-derived compounds trans-cinnamaldehyde and eugenol. *Applied and Environmental Microbiology, 78*, 2981-2987.

Kollanoor-Johny, A., Mattson, T., Baskaran, S. A., Amalaradjou, M., Hoagland, T. A., Darre, M. J., Khan, M. I. (2012b). Caprylic acid reduces *Salmonella* Enteritidis populations in various segments of digestive tract and internal organs of 3- and 6-week-old broiler chickens, therapeutically. *Poultry Science, 91*, 1686-1694.

Kurita, A., Gotoh, H., Eguchi, M., Okada, N., Matsuura, S., Matsui, H., Danbara, H. (2003). Intracellular expression of the *Salmonella* plasmid virulence protein, *SpvB*, causes apoptotic cell death in eukaryotic cells. *Microbial Pathogenesis 35*, 43–48.

Lesnick, M.L., Reiner, N.E., Fierer, J., & Guiney, D.G. (2001). The *Salmonella spvB* virulence gene encodes an enzyme that ADP-ribosylates actin and destabilizes the cytoskeleton of eukaryotic cells. *Moleculur Microbiology 39*, 1464–1470.

Leverentz, B., Conway, W. S., Alavidze, Z., Janisiewicz, W. J., Fuchs, Y., Camp, M. J., Chighladze, E., & Sulakvelidze, A. (2001). Examination of bacteriophage as a biocontrol method for *Salmonella* on fresh-cut fruit: a model study. *Journal Food Protection, 64*, 1116–1121.

Ley, R. E., Turnbaugh, P. J., Klein, S., & Gordon, J. I. (2006). Microbial ecology: human gut microbes associated with obesity. *Nature, 444*, 1022-1023.

Li, H., Xu, H., Zhou, Y., Zhang, J., Long, C., Li, S., Chen, S. (2007). The phosphothreonine lyase activity of a bacterial type III effector family. *Science, 315*, 1000–1003.

Liljebjelke, K. A., Hofacre, C. L., Liu, T., White, D. G., Ayers, S., Young, S., & Maurer, J. J. (2005). Vertical and horizontal transmission of Salmonella within integrated broiler production system. *Foodborne Pathogens & Disease*, *2*, 90-102.

Lynne, A. M., Kaldhone, P., David, D., White, D. G., & Foley, S. L. (2009). Characterization of antimicrobial resistance in *Salmonella enterica* serotype Heidelberg isolated from food animals. *Foodborne pathogens and disease*, *6*, 207-215.

Mani-López, E., García, H. S., & López-Malo, A. (2012). Organic acids as antimicrobials to control *Salmonella* in meat and poultry products. *Food Research International*, *45*(2), 713-721.

Matulova, M., Havlickova, H., Sisak, F., & Rychlik, I. (2012). Vaccination of chickens with *Salmonella* pathogenicity island (SPI) 1 and SPI2 defective mutants of *Salmonella enterica* serovar enteritidis. *Vaccine, 30*, 2090-2097.

Mazurek, J., Salehi, E., Propes, D., Holt, J., Bannerman, T., Nicholson, L.M., Bundesen, M., Duffy, R., & Moolenaar, R.L. (2004). A multistate outbreak of *Salmonella enterica* serotype Typhimurium infection linked to raw milk consumption—Ohio, 2003. *Journal of Food Protection, 67,* 2165–2170.

Mazurkiewicz, P., Thomas, J., Thompson, J.A., Liu, M., Arbibe, L., Sansonetti, P., and Holden, D.W. (2008) SpvC is a *Salmonella* effector with phosphothreonine lyase activity on host mitogen-activated protein kinases. *Molecular Microbiology, 67*, 1371–1383.

McCarthy, N., Reen, F. J., Buckley, J. F., Frye, J. G., Boyd, E. F., & Gilroy, D. (2009). Sensitive and rapid molecular detection assays for *Salmonella enterica* serovars Typhimurium and Heidelberg. *Journal of Food Protection, 11,* 2236-2435.

Medalla, F., Hoekstra, R. M., Whichard, J. M., Barzilay, E. J., Chiller, T. M., Joyce, K., Rickert, R. (2013). Increase in resistance to Ceftriaxone and nonsusceptibility to Ciprofloxacin and decrease in multidrug resistance among *Salmonella* strains, United States, 1996-2009. *Foodborne Pathogens and Disease, 10*, 302-309.

Menconi, A., Wolfendum, A. D., Shivaramaiah, S., Terraes, J. C., Urbano, T., Kuttel, J., Kremer, C. (2011). Effect of lactic acid bacteria probiotic culture for the treatment of *Salmonella enterica* serovar Heidelberg in neonatal broiler chickens and turkey poults. *Poultry Science, 90*, 561-565.

Methner, U., Rammler, N., Fehlhaber, K., & Rosler, U. (2011). *Salmonella* status of pigs at slaughter — bacteriological and serological analysis. *International Journal of Food Microbiology, 151* (1), 15-20.

Modi, R., Hirvi, Y., Hill, A., & Griffiths, M. W. (2001). Effect of phage on survival of *Salmonella* Enteritidis during manufacture and storage of cheddar cheese made from raw and pasteurized milk. *Journal Food Protection, 64*, 927–933.

Morishita T. Y., Aye, P. P., Harr, B S., Cobb, C. W., & Clifford, J. R. (1997). Evaluation of an avian-specific probiotic to reduce the colonization and shedding of *Campylobacter jejuni* in broilers. *Avian Diseases, 41*, 850-855.

Nandre, R. M., Katsuda, K., Chaudhari, A. A., Kim, B., & Lee, J. H. (2012). A genetically engineered derivative of *Salmonella* Enteritidis as a novel live vaccine candidate for salmonellosis in chickens. *Research in Veterinary Science, 93*, 596-603.

National Antimicrobial Resistance Monitoring System (NARMS). (2010). 2010 Executive Report. Available at http://www.fda.gov/downloads/AnimalVeterinary/SafetyHealth/AntimicrobialResistance/NationalAntimicrobialResistanceMonitoringSystem/UCM312360.pdf. Accessed on 22 september 2014.

Nayak, R., Call, V., Kaldhone, P., Tyler, C., Anderson, G., Phillips, S., & Foley, S. L. (2007). Comparison of *Salmonella enterica* serovar Heidelberg susceptibility testing results. *Clinical Medicine & Research, 5*, 98-105.

Newman, D. J. & Cragg, G. M. (2012). Natural products as sources of new drugs over the 30 years from 1981 to 2010. *Journal of Natural Products, 75*, 311-335.

Newman, D. J., Cragg, G. M., & Snader, K. M. (2000). The influence of natural products upon drug development. *Natural Product Reports, 17,* 215-234.

Ohl, M. E. & Miller, S. I. (2001). *Salmonella:* A model for bacterial pathogenesis. *Annual Review of Medicine, 52*, 259-274.

Ohno, T., Kita, M., Yamaoka, Y., Imamura, S., Yamamoto, T., Mitsufuji, S., Mitsufuji, S., Kodama, T., Kashima, K. & Imanishi, J. (2003). Antimicrobial activity of essential oils against *Helicobacter pylori. Helicobacter, 8,* 207-215.

Olszak, T., An, D., Zeissig, S., Vera, M. P., Richter, J. Franke, A., Glickman, J. N. (2012). Microbial exposure during early life has persistent effects on natural killer T-cell function. *Science, 336*, 489-493.

Ordonez, G., Llopis, N., & Penalver, P. (2008). Efficacy of eugenol against a *Salmonella enterica* serovar Enteritidis experimental infection in commercial layers in production. *Journal of Applied Poultry Research, 17*, 376-382.

Patchanee, P., Zewde, B. M., Tadesse, D. A., Hoet, A., & Gebreyes, W. A. (2008). Characterization of multidrug-resistant *Salmonella enterica* serovar Heidelberg isolated from humans and animals. *Foodborne pathogens and disease*, *5*, 839-851.

Pathania, A., McKee, S. R., Bilgili, S. F., & Singh, M. (2010). Antimicrobial activity of commercial marinades against multiple strains of *Salmonella* spp. *International journal of food microbiology*, *139*, 214-217.

Piddock, L. J. V. (2002). Fluoroquinolone resistance in *Salmonella* serovars isolated from humans and food animals. *FEMS Microbiology Reviews, 26,* 3-16.

Piddock, L. J. V., Ricci, V., McLaren, I., & Griggs, D. J. (1998). Role of mutations in the *gyrA* and *parC* genes of 196 nalidixic-acid-resistant *Salmonella* serotypes isolated from animals. *The Journal of Antimicrobial Chemotherapy, 41,* 635-642.

Poppe, C., Johnson, R. P., Forsberg, C. M., & Irwin, R. J. (1992). *Salmonella* Enteritidis and other *Salmonella* in laying hens and eggs from flocks with Salmonella in their environment. *Canadian Journal of Veterinary Research*, *56*, 226.

Promsopone, B., Morishita, T. Y., Aye, P. P., Cobb, C. W., Veldkamp, A., & Clifford, J. R. (1998). Evaluation of an avian-specific probiotic and *Salmonella* Typhimurium-specific antibodies on the colonization of *Salmonella* Typhimurium in broilers. *Journal of Food Protection, 2*, 141-256.

Qu, A., Brulc, J. M., Wilson, M. K., Law, B. F., Theoret, J. R., Joens, L. A., Konkel, M. E. (2008). Comparative metagenomics reveals host specific metavirulomes and horizontal gene transfer elements in the chicken cecum microbiome. *PLoS ONE, 3*, 2945.

Reichling, J. (2010). Annual plant reviews: Plant-microbe interactions and secondary metabolites with antibacterial, antifungal and antiviral properties. Wink, M. (Ed). Functions and biotechnology of plant secondary metabolites, second edition. Oxford, UK:Wiley-Blackwell.

Revolledo, L., Ferreira, C. S. A., & Ferreira, A. J. P. (2008). Prevention of *Salmonella* Typhimurium colonization and organ invasion by combination treatment in broiler chicks. *Poultry Science*, *88*, 734-743.

Ross, R. P., Morgan, S., & Hill, C. (2002). Preservation and fermentation: past, present and future. *International Journal of Food Microbiology*, 79, 3-16.

Scallan, E., Griffin, P. M., Angulo, F. J., Tauxe, R. V., & Hoekstra, R. M. (2011). Foodborne illness acquired in the United States—unspecified agents. *Emerging infectious diseases*, *17*, 16.

Schmitt, C.K., Ikeda, J.S., Darnell, S.C., Watson, P.R., Bispham, J., Wallis, T.S. (2001). Absence of all components of the flagellar export and synthesis machinery differentially alters virulence of *Salmonella enterica* serovar Typhimurium in models of typhoid fever, survival in macrophages, tissue culture invasiveness, and calf enterocolitis. *Infection and Immunity, 69*, 5619–5625.

Sefton, A. M. (2002). Mechanisms of antimicrobial resistance: Their clinical relevance in the new millennium. *Drugs, 62*, 557-566.

Sheppard, M., Webb, C., Heath, F., Mallows, V., Emilianus, R., Maskell, D., & Mastroeni, P. (2003). Dynamics of bacterial growth and distribution within the liver during *Salmonella* infection. *Cellular Microbiology, 5*, 593-600.

Skurnik, M. & Strauch, E. (2006). Phage therapy: facts and fiction. *International Journal of Medical Microbiology, 296* (1), 5-14.

Sonnenburg, J. L., Chen, C. T. L., & Gordon, J. (2006). Genomic and metabolic studies of the impact of probiotics on a model gut symbiont and host. *PLoS Biology, 4* (12), e413, doi:10.1371/journal.pbio.0040413.

Tellez, G., Petrone, V. M., Escorcia, M., Morishita, T. Y., Cobb, C. W., & Villasenõr, L. (2001). Evaluation of avian-specific probiotic and *Salmonella* Enteritidis-, *Salmonella* Typhimurium-, and *Salmonella* Heidelberg- specific antibodies on cecal colonization and organ invasion of *Salmonella* enteritidis in broilers. *Journal of Food Protection, 3*, 287-291.

Tezcan-Merdol, D., Nyman, T., Lindberg, U., Haag, F., Koch-Nolte, F., & Rhen, M. (2001). Actin is ADP-ribosylated by the *Salmonella enterica* virulence-associated protein SpvB. *Molecular Microbiology, 39*, 606–619.

Torok, V. A., Allison, G. E., Percy, N. J., Ophel-Keller, K., & Hughes, R. (2011). Influence of antimicrobial feed additives on broiler commensal posthatch gut microbiota development and performance. *Applied Environmental Microbiology, 77*, 3380-3390.

Upadhyaya, I., Upadhyay, A., Kollanoor-Johny, A., Baskaran, S. A., Mooyottu, S., Darre, M. J., & Venkitanarayanan, K. (2013). Rapid inactivation of *Salmonella* Enteritidis on shell eggs by plant-derived antimicrobials. *Poultry Science, 92*, 3228-3235.

US Department of Agriculture, Economic Research Service. (2010). Foodborne illness cost calculator: Salmonella. Washington, DC: US Department of Agriculture, Economic Research Service; 2010. Available at http://www.ers.usda.gov/data/foodborneillness/salm_intro.asp. Accessed September 5, 2014.

Valle, J., Da Re, S., Henry, N., Fontaine, T., Balestrino, D., Latour-Lambert, & Ghigo, J. (2006). Broad-spectrum biofilm inhibition by a secreted bacterial polysaccharide. *Proceedings of the National Academy of Sciences, 103*, 12558-12563.

Varma, J. K., Molbak, K., Barrett, T. J., Beebe, J. L., Jones, T. F., Rabatsky-Her, T., Smith, K. E. (2005). Antimicrobial-resistant nontyphoidal *Salmonella* is associated with excess bloodstream infections and hospitalizations. *Journal of Infectious Diseases, 191*, 554-561.

Vicente, J. L., Torres-Rodriguez A., Higgins S. E., Pixley C., Tellez G., Donoghue A. D., & Hargis, B. M. (2008). Effect of a selected lactobacillus spp.–based probiotic on *Salmonella enterica* serovar enteritidis–infected broiler chicks. *Avian Diseases, 52*, 143-146.

Vugia, D. J., Samuel, M., Farley, M. M., Marcus, R., Shiferaw, B., Shallow, S., Smith, K. (2004). Invasive *Salmonella* infections in the United States, FoodNet, 1996-1999: incidence, serotype distribution, and outcome. *Clinical Infectious Diseases, 15*, S149-156.

Waters, C. M. & Bassler, B. L. (2005). Quorum sensing: cell-to-cell communication in bacteria. *Cell and Development Biology*, *21*, 319-346.

Wilmshurst, P., & Sutcliffe, H. (1995). Splenic abscess due to Salmonella Heidelberg. *Clinical infectious diseases*, *21*, 1065-1066.

Yatsunenko, T., Rey, F. E., Manary, M. J., Trehan, I., Dominguez-Bello, M. G., Contreras, M., Magris, M. (2012). Human gut microbiome viewed across age and geography. *Nature, 486,* 222-227.

Zhao, S., White, D. G., Friedman, S. L., Glenn, A., Blickenstaff, K., Ayers, S. L., & McDermott, P. F. (2008). Antimicrobial resistance in Salmonella enterica serovar Heidelberg isolates from retail meats, including poultry, from 2002 to 2006. *Applied and environmental microbiology*, *74*, 6656-6662.

Zou, W., Al-Khaldi, S. F., Branham, W. S., Han, T., Fuscoe, J. C., Han, J., ... & Nayak, R. (2010). Microarray analysis of virulence gene profiles in Salmonella serovars from food/food animal environment. *The Journal of Infection in Developing Countries*, *5*, 94-105.

In: Salmonella
Editor: Christopher B. Hackett
ISBN: 978-1-63463-651-3

Chapter 3

USE OF NATURAL ANTIMICROBIALS AS A TREATMENT OPTION TO CONTROL *SALMONELLA* TYPHIMURIUM

Maria Sanz-Puig, M. C. Pina-Pérez, Antonio Martínez and Dolores Rodrigo*
Instituto de Agroquímica y Tecnología de Alimentos (IATA-CSIC), Valencia, Spain

ABSTRACT

Salmonella is a foodborne pathogen that causes a huge amount of cases of typhoid fever, gastroenteritis, and deaths every year throughout the world. Although salmonellosis cases in humans have decreased in the last five years, *Salmonella* remains the second most common zoonosis in humans. Foodborne outbreaks caused by *Salmonella* have also reduced in recent years, but they have been linked with contamination of eggs and egg products, cheese, mixed foods, and fresh fruits and vegetables. Therefore control measures for this microorganism are very important to prevent and control *Salmonella* at relevant stages of production, processing, and distribution, especially in primary production, thus reducing its prevalence and the risk it poses to public health. In this context, research carried out to find antimicrobial compounds from natural sources is important because they could be used as additives in new product formulations, where they could exercise an additional measure to control *Salmonella* growth and have an important impact from economic and food safety points of view.

By-products from the food industry are a potential source of inexpensive raw materials, and are rich in bioactive components whose technological and antimicrobial properties are still scarcely studied. With the aim of covering this gap, the objective of the present study was focused on evaluating the antimicrobial properties of three citrus by-products – mandarin, orange, and lemon – against *Salmonella enterica* serovar

* Corresponding author: Dolores Rodrigo. Instituto de Agroquímica y Tecnología de Alimentos (IATA-CSIC). Avd. Agustín Escardino, 7, 46980 Paterna (Valencia) Spain. E-mail: lolesra@iata.csic.es, phone: +34 963900022, fax: +34 963636301.

Typhimurium, in reference medium, under various incubation conditions with differences in temperature and by-product concentration. According to the results obtained, it can be concluded that all the citrus by-products showed a bacteriostatic and/or bactericidal effect under the conditions studied, the mandarin by-product being the most effective one. Maximum reduction levels in the microbial population attained values of ≈8 $\log_{10}$ cycles at refrigeration temperature (5 °C). Consequently, it can be concluded that citrus by-products have effective antimicrobial activity, and could act as an additional barrier to microbial growth when added to pasteurized beverages that are stored under refrigeration, contributing additionally to meeting the zero waste targets set by the European Union.

1. *Salmonella*: A Foodborne Pathogen

Salmonella is one of the most important foodborne pathogens worldwide, producing an illness called salmonellosis that causes over 90,000 human cases per year in the European Union. Salmonellosis is a zoonotic disease that can be transmitted between animals and humans directly or indirectly, and it usually produces diarrhea, nausea, fever, and abdominal cramps, although if it infects the bloodstream it can be life-threatening (EFSA, 2014).

Salmonella is usually present in the intestines of birds and mammals and can be transferred to humans through contaminated foods such as eggs and raw meat from pigs, turkeys, and chickens. The incubation period ranges from five hours to seven days, but the clinical signs usually appear 12 to 16 hours after ingestion of contaminated food and the syndrome lasts between two and seven days. Usually infections occur in people at risk, young, elderly, or immunocompromised people (Forshell and Wierup, 2006).

In the same way as with humans, *Salmonella* infects animals too. There are serovars of *Salmonella* that are adapted to specific animal species, such as *S.* Abortus ovis (sheep), *S.* Cholerae suis (pigs), *S.* Gallinarum (poultry), *S.* Abortus equi (horses), and *S.* Dublin (cattle). These serovars are not pathogenic to humans, but if humans are infected these serovars cause septicemia. In contrast, these host-adapted serovars cause abortions and severe gastroenteritis in their animal hosts (EU, 2003).

Among *Salmonella* species, there is a group consisting of *S.* Typhimurium, *S.* Enteritidis, *S.* Hadar, *S.* Infantis, and others, which infect both humans and animals. They can establish an animal infection without clinical signs during a variable time period, which can produce a potential zoonosis. Also, serovars that are usually non-pathogenic can cause disease in animal species used for food products under stress conditions (Forshell and Wierup, 2006).

Salmonella zoonosis can be transmitted from various animal sources. The food categories with the highest hazard in relation to zoonosis are raw meat, raw and undercooked poultry meat, eggs and their derivative products, unpasteurized milk and its derivative products, sprouted seeds, unpasteurized fruit juices, and home-made mayonnaise.

Therefore, *Salmonella* control measures are very important to guarantee human health. In this connection, since 1980, when the WHO formulated a three-point defense strategy against *Salmonella* (WHO, 1980), the following measures have been carried out:

- Pre-harvest control: Control of *Salmonella* in food-producing animals. Establishment of monitoring programs to find and control sources of *Salmonella* infection and prevent further outbreaks, with the aim of producing *Salmonella*-free animals.

- Harvest control: Guarantee hygiene during slaughter and processing of meat and meat products.
- Post-harvest control: Educate both the food industry and consumers about good hygiene practices.

Pursuing the same goal, in 2003 the European Union set up control measures to combat zoonosis, considering *Salmonella* as a priority because of the high number of cases of salmonellosis every year and their economic cost. In this connection, several programs to control *Salmonella* have been implemented in all Member States of the European Union.

In these programs, EFSA provides recommendations for control and reduction measures, with the aim of supporting the reduction of *Salmonella* in the food chain (reduction targets in poultry flocks and poultry meat, use of vaccines and antimicrobials to control *Salmonella*). Also, EFSA conducted studies on the prevalence of *Salmonella* in food and food-producing animals and evaluation of the risk factors that affect its prevalence in animals and food.

The application of these programs and the coordinated efforts made by all EU members have resulted in a significant reduction of human cases of *Salmonella* amounting to almost 50% in 5 years (2004–2009). The prevalence of *Salmonella* in flocks of laying hens has also been reduced to 2% or less in all EU Member States, from original values of 20% in some of them. The main reason for the decrease in *Salmonella* cases in humans is probably the reduction of these bacteria in laying hen flocks, because eggs are the most important source of human infections in the EU (EFSA, 2012).

However, although salmonellosis cases in humans have decreased in the last five years, *Salmonella* remains the second most common zoonosis in humans, with almost 200,000 reported human cases in 2012. Therefore measures for the control and inactivation of this microorganism are very important to prevent, detect, and control *Salmonella* at relevant stages of production, processing, and distribution, especially in primary production, to reduce its prevalence and the risk it poses to public health.

2. *Salmonella* Control Measures

Salmonella has an important effect on foodborne illnesses; therefore control measures against this microorganism are necessary.

Traditionally, antimicrobial drugs such as antibiotics or chemical substances have been used to control *Salmonella* spp., among other foodborne pathogens. However, the development of resistance to these antimicrobial agents by the microorganisms and public concern about the health damage caused by synthetic additives have led consumers to reject chemical preservatives in food products. Therefore processors and scientists are working together to find antimicrobial compounds of natural origin. In this connection, many extracts of plant origin have proved to have a large range of bioactive compounds with antioxidant, anti-carcinogenic, anti-inflammatory, or antimicrobial properties, including polyphenol compounds and essential oils from various plants (Dai and Mumper, 2010), such as clove and cinnamon (Cava et al., 2007), olives (Ferrer et al., 2009), brassicas (Brandi et al., 2006), and *Stevia rebaudiana* (Belda-Galbis et al., 2014), among others.

The bioactive properties of these natural compounds can be considered as natural additives in the development of new products with the goal of eliminating chemical additives in our food (Khan et al., 2014).

Essential oils are secondary plant metabolites that exert a potential antimicrobial effect to control foodborne pathogenic bacteria, such as *Salmonella* species, owing to the presence of bioactive volatile compounds. Although their antimicrobial capacity depends on several factors, such as temperature, pH, microbial population load, and favorable environment, they are potent antimicrobials with a low toxic effect, and this makes them green preservatives for microbial control in the food industry. Consequently, the use of essential oils has become an important area of research for future applications as pathogen control measures both in human food systems and in animal nutrition (Bajpai et al., 2012; Palaniappan and Holley, 2010; Losa, 2001; Borsoi et al., 2011). Many research studies have demonstrated their antimicrobial effect against *Salmonella* in different food models, such as the effect shown by oregano in salads, meat products, and tomatoes (Koutsoumanis et al., 1999; Govaris et al., 2010; Gunduz et al., 2010b), the effect of cinnamon in liquid whole egg (Valverde et al., 2010), or the effect of citral in fishery products (Kim et al., 1995b) against *Salmonella*.

The bioactive capacity of essential oils is generally attributed to their chemical compounds, such as polyphenolic or terpene groups. Many studies have shown the antimicrobial effect of various polyphenol plant extracts against *Salmonella* (Bajpai et al., 2012). For example, the polyphenol content of raspberry, cloudberry, and strawberry (species of the *Rubus* and *Fragaria* genera) showed an antibacterial effect against several strains of *Salmonella* (Puupponen-Pimia et al., 2001); Karapinar et al. (2007) showed that koruk (unripe grape from *Vitis vinifera*) juice is effective to inhibit *S.* Typhimurium; and Fattouch et al. (2007) showed that Tunisian quince (*Cydonia oblonga* Miller) pulp and peel polyphenolic extracts exerted antimicrobial activity against *Salmonella* spp. isolated from food.

Moreover, these natural bioactive compounds from plants can also be extracted from food industry waste, which generally consists of peel, seeds, and outer leaves of fruits and vegetables that remain after food processing (Marín et al., 2007; O'Shea et al., 2012), and that are usually consigned to landfill or incineration and represent a cost and an environmental problem for the companies involved.

However, these apparently valueless wastes can be considered as a renewable source of raw material, and their use can have a double benefit: reducing pollution and turning them into substances with added value (Martín-Luengo et al., 2011; Schieber et al., 2001).

Moreover, agri-food by-product valorization is a requirement of the European Union (EUROSTAT, 2010), and many research studies nowadays are focused on recovering, revaluing, and recycling these by-products.

Various ways of using these by-products have been developed. They can be used in agriculture as phytochemical compounds, in waste water treatment as biosorbents, in feed production, in the paper industry, as fuel, or as additives in the development of new products (Gracia, 2004).

An important percentage of total by-product production worldwide consists of citrus by-products. Citrus is the largest fruit crop worldwide, with 100 million tons of annual production, mainly from Brazil, the US, and Mediterranean countries (Djilas, 2009). The industrial production of marmalade and citrus segments and the extraction of flavonoids and essential oils result in 15 million tons of citrus waste a year throughout the world.

Many research studies have shown the health benefits of bioactive compounds that have been found in citrus by-products, mainly phenolic compounds such as carotenoids and flavonols (Sawalha et al., 2009; Ghafar et al., 2009; Igual et al., 2013). Consequently, the use of citrus by-products as functional ingredients in the development of new food products is a promising possibility.

Therefore the aim of this research study is to evaluate the antimicrobial effect of mandarin, orange, and lemon by-products against *Salmonella enterica* serovar Typhimurium under various incubation conditions.

3. Use of Natural Antimicrobials against *Salmonella*

In this work, the antimicrobial effect of three citrus by-products (mandarin, orange, and lemon) against *S.* Typhimurium was evaluated at different conditions, with incubation temperatures in the range [5–22] °C and various citrus by-product concentrations (0, 0.5, 1, 5, 10%), in reference medium (buffered peptone water (1‰ (w/v))).

The decimal log cycles of *S.* Typhimurium inactivation under the conditions studied, after 96 hours of incubation at 5 and 10 °C and 24 hours at 22 °C, are shown in Tables 1 (mandarin), 2 (orange), and 3 (lemon). As can be seen in these tables, the three citrus by-products that were tested showed an antimicrobial effect against *S.* Typhimurium. Mandarin was the by-product with the best antimicrobial effect, achieving a maximum inactivation level of approximately 8 log cycles at refrigeration temperature (5 °C) with 5% mandarin by-product addition. Moreover, at 10 °C and 22 °C mandarin by-product also had a bacteriostatic effect, with maximum values of 8 and 2 log cycles of microbial inactivation, respectively.

Mandarin by-product effectiveness was followed by orange by-product, which achieved a maximum of 3.59 log cycles of microbial inactivation, also at refrigeration temperature (5 °C), with 10% orange addition. At 10 °C, orange by-product also had a bactericidal effect, with a maximum of 1.5 log cycles of microbial inactivation, and at 22 °C it was bacteriostatic.

Finally, although lemon by-product was able to inhibit *S.* Typhimurium growth at all the concentrations and temperatures studied, it was the natural extract that showed the smallest antimicrobial effect, with a maximum inactivation level of 1.22 log cycles when 10% of lemon by-product was added at 22 °C.

If we compare the data in the three tables, it can be seen that the higher the citrus by-product concentration, the greater the *S.* Typhimurium inactivation level achieved by the three by-products under study. An ANOVA analysis confirmed that the citrus by-product concentration had a significant influence ($p \leq 0.05$) on the antimicrobial effect observed against *S. Typhimurium*, although there were no significant differences between 5 and 10% by-product addition.

Regarding the effect of temperature, the tables show that microbial inactivation levels were generally higher at lower temperatures. An ANOVA analysis confirmed that temperature had a significant influence ($p \leq 0.05$) on the antimicrobial effect observed against *S.* Typhimurium. Therefore refrigeration temperature also showed an antimicrobial effect against *S.* Typhimurium.

Figure 1 shows the evolution of the microbial load during the incubation period, under exposure to 5% (w/v) of each of the citrus by-products, at 5, 10, and 22 °C.

Table 1. Inactivation levels (log cycles) of *S.* Typhimurium under exposure to mandarin by-product at different conditions of temperature and by-product concentration

Mandarin by-product concentration	*S.* Typhimurium inactivation levels		
	5 °C	10 °C	22 °C
0%	-0.1761	0.2404	1.5117
0.5%	-2.5539	0.2032	-0.8391
1%	-3.6193	0.1123	-1.0591
5%	-7.9243	-7.5798	-2.0838
10%	-7.8865	-7.8921	-1.8792

Table 2. Inactivation levels (log cycles) of *S.* Typhimurium under exposure to orange by-product at different conditions of temperature and by-product concentration

Orange by-product concentration	*S.* Typhimurium inactivation levels		
	5 °C	10 °C	22 °C
0%	-0.1761	0.2404	1.5117
0.5%	-0.1231	-1.5607	0.0570
1%	-0.4506	-1.3912	0.0536
5%	-3.4610	-1.4993	-1.0269
10%	-3.5911	-1.5586	-1.0280

Table 3. Inactivation levels (log cycles) of *S.* Typhimurium under exposure to lemon by-product at different conditions of temperature and by-product concentration

Lemon by-product concentration	*S.* Typhimurium inactivation levels		
	5 °C	10 °C	22 °C
0%	-0.1761	0.2404	1.5117
0.5%	-0.1231	-0.1845	0.0570
1%	-0.4506	-0.0209	0.0536
5%	-0.8329	-0.5616	-1.1622
10%	-0.9945	-0.5991	-1.2013

If we compare these results with the control sample (buffered peptone water without citrus by-product addition), we can see that at refrigeration temperature (5 °C) *S.* Typhimurium growth was inhibited even in the control sample, as also occurred in other research studies (Yang et al., 2001; Mañas et al., 2003).

However, at the same temperature, the addition of 5% of mandarin or orange by-product had a bactericidal effect, achieving a maximum of 8 log cycles of *S.* Typhimurium inactivation in the case of mandarin.

In contrast, at 10 °C *S.* Typhimurium started to grow in control samples, but the addition of 5% of mandarin and orange by-product showed a bactericidal effect, again achieving a maximum of 8 log cycles of microbial inactivation by mandarin by-product. The addition of lemon by-product had a bacteriostatic effect.

At 22 °C, the initial population of *S.* Typhimurium started to grow, but addition of 5% (w/v) of the three citrus by-products under study had an inhibitory effect on microbial growth.

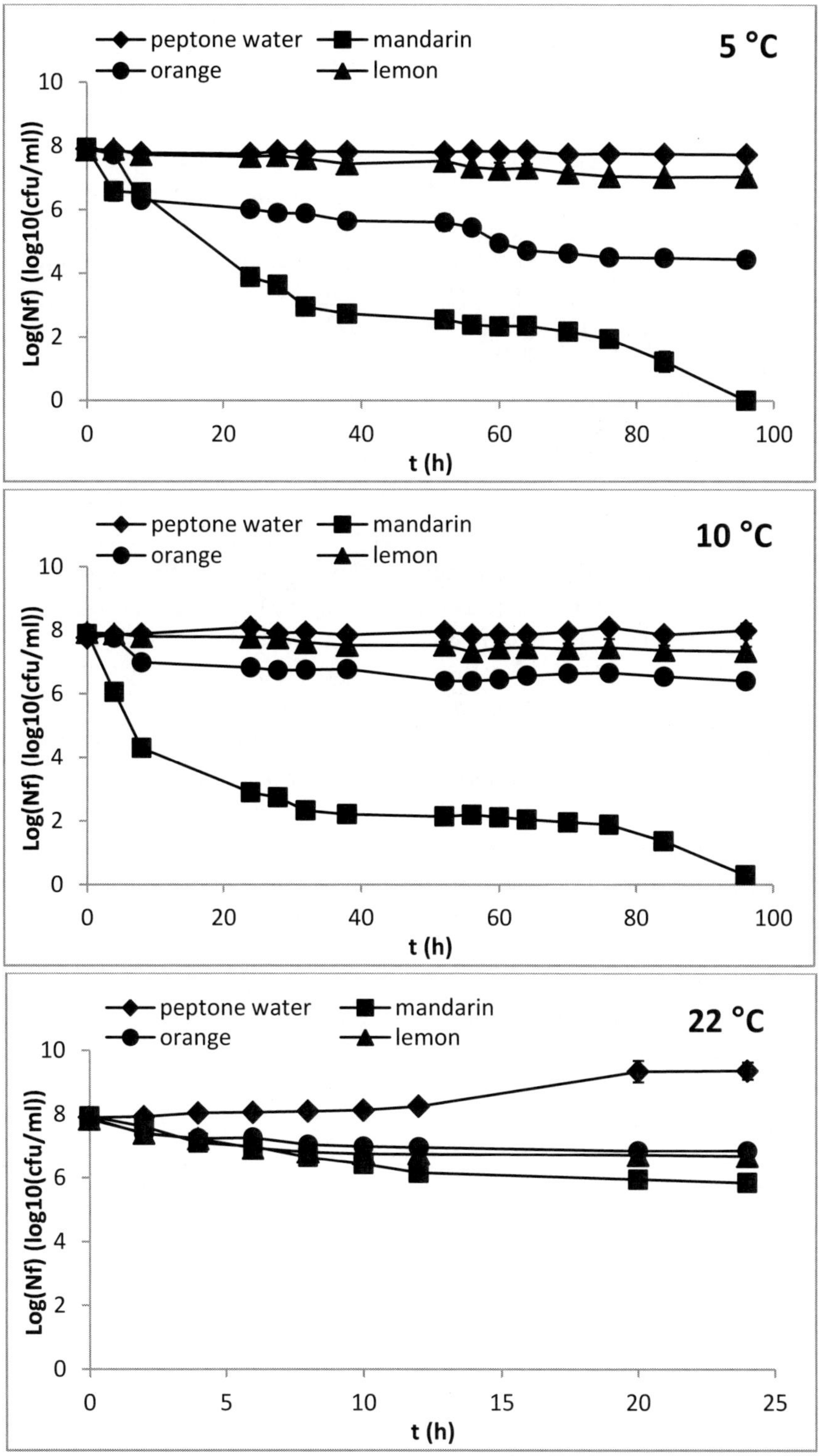

Figure 1. Evolution of initial *S.* Typhimurium cell population under the effect of mandarin, orange, and lemon by-product at 5% at 5, 10, and 22 °C.

The bacteriostatic effect of refrigeration temperature has also been extensively recognized by various authors, who attributed this effect to a stress response mechanism of microorganisms, due to molecular changes and metabolic defense mechanisms (Shapiro and Cowen, 2012). Similarly, this control measure can help to enhance the effectiveness of other thermal and non-thermal technologies, especially during the shelf life of the product, achieving additive or synergistic effects in the reduction of the bacterial load until consumption. In the present study, the combination of the two factors, refrigeration temperature and citrus by-product concentration in the medium, showed a synergistic effect regarding microbial inactivation at 5 and 10 °C. The same results were observed by De Oliveira et al. (2013), who showed the antimicrobial effect of oregano and lemongrass essential oils against *Salmonella* Enteritidis in ground beef at refrigeration temperature.

In addition to temperature and by-product concentration, there are other factors that can influence *S.* Typhimurium growth, such as pH and citrus by-product composition.

In acid products, pH has an important control effect on microbial growth (Alali et al., 2012). However, the pH values of the three citrus by-products are in the range of 3.77 to 4.54, with the lemon by-product having the lowest pH value, followed by the orange and mandarin by-products, respectively. On the other hand, the citrus by-product that showed the highest antimicrobial capacity was mandarin. Thus there was no concordance between the pH value and the antimicrobial effect of the by-products under study against *S.* Typhimurium. Consequently, in this case pH was not the most important factor that contributed to the capacity of these citrus by-products to inhibit microbial growth.

Regarding the citrus by-product composition, many studies show the antimicrobial properties of bioactive compounds of citrus peels and seeds, which mainly belong to the polyphenol group (Espina et al., 2011; Viuda-Martos et al., 2008). They are lipidic compounds with aromatic properties, and they also have an antimicrobial effect that is of interest for the pharmaceutical and food industries (Sobrino-López et al., 2006). The total polyphenol content of the three citrus by-products under study was in the range of 4600 to 5111 mg gallic acid/L, with the mandarin by-product having the highest value, followed by orange and lemon, respectively. Therefore in this case it is possible to establish a concordance between the antimicrobial effect of the citrus by-products and their polyphenol content. The mechanism of action of these compounds is still not well understood, but the most accepted hypothesis is that their hydrophobic components can break down the lipid components of the bacterial membrane and then the cell content is released to the exterior (Trípoli et al., 2007).

In conclusion, all the citrus by-products under study showed a bactericidal and bacteriostatic capacity against *S.* Typhimurium, with the mandarin by-product having the best antimicrobial capacity, especially at refrigeration temperature. Their demonstrated beneficial properties, both nutritional and bioactive, make them possible candidates to be added to food products for both animals and humans as a microbial control measure.

The bactericidal and bacteriostatic capacity that they demonstrated suggests the possibility of using them as natural bacteriostatic compounds on crops as a measure to control the growth of foodborne pathogens, in liquid form on vegetables and cereals and as wax on fruit peel.

Furthermore, in view of their capacity to inactivate zoonotic microorganisms such as *S.* Typhimurium, they might play an important role in control of zoonotic cases if they are added to the feed of animals for human consumption.

On the other hand, their antimicrobial effect against *S.* Typhimurium, especially at refrigeration temperature, opens the door to the possibility of using them as ingredients in food products for humans that are subjected to pasteurization treatment and subsequently stored at refrigeration temperature, as a control measure against foodborne pathogens such as *S.* Typhimurium. Thus they might be a possible solution in the increasing search by food producers for new products with added value, in response to increasing consumer demand for natural products with health benefits (O'Shea et al., 2012).

Consequently, agri-food by-products could be re-valorized as antimicrobial additives and would no longer be an economic problem; on the contrary, the valorization of these natural compounds could represent an economic benefit for food companies, adding nutritional and antimicrobial potential to newly developed products.

REFERENCES

Alali, W. Q., Mann, D. A. and Beuchat, L. R. (2012). Viability of *Salmonella* and *Listeria monocytogenes* in delicatessen salads and hummus as affected by sodium content and storage temperature. *Journal of Food Protection*, 75(6): 1043-1056.

Bajpai, V. K., Baek, K. H. and Kang, S. C. (2012). Control of *Salmonella* in foods by using essential oils: A review. *Food Research International*, 45, 722-734.

Belda-Galbis, C. M., Pina-Pérez, M. C., Espinosa, J., Marco-Celdrán, A., Martínez, A., and Rodrigo, D. (2014). Use of the modified Gompertz equation to assess the *Stevia rebaudiana* Bertoni antilisterial kinetics. *Food Microbiology*, 38, 56-61.

Borsoi, A., Santos, L. R., Diniz, G. S., Salle, C. T. P., Moraes, H. L. S., and Nascimiento, V. P. (2011). *Salmonella* Fecal Excretion Control in Broiler Chickens by Organic Acids and Essential Oils Blend Feed Added. *Brazilian Journal of Poultry Science*, 13, 65-69.

Brandi, G., Amagliani, G., Schiavano, G. F., De Santi, M., and Sisti, M. (2006). Activity of *Brassica oleracea* Leaf Juice on Foodborne Pathogenic Bacteria. *Journal of Food Protection*, 69, 9, 2274-2279.

Cava, R., Nowak, E., Taboada, A., and Marin-Iniesta, F. (2007). Antimicrobial Activity of Clove and Cinnamon Essential Oils against *Listeria monocytogenes* in Pasteurized Milk. *Journal of Food Protection*, 70, 12, 2757-2763.

Dai, J. and Mumper, R. J. (2010). Plant Phenolics: Extraction, Analysis and their Antioxidant and Anticancer Properties. *Molecules*, 15, 7313-7352.

De Oliveira, T. L. C., Soares, R. A. and Piccoli, R. H. (2013). A Weibull model to describe antimicrobial kinetics of oregano and lemongrass essential oils against *Salmonella* Enteritidis in ground beef during refrigerated storage. *Meat Science*, 93, 3, 645-651.

Djilas, S. (2009). By-products of fruits processing as a source of phytochemicals. *Chemical Industry and Chemical Engineering Quarterly*, 15 (4) 191-202.

EFSA. (2014). EFSA explains zoonotic diseases: *Salmonella.*

EFSA. (2012). EU measures help reduce human *Salmonella* cases by almost one-half. 30 January 2012.

Espina, L., Somolinos, M., Lorán, S., Conchello, P., García, D., and Pagán, R. (2011). Chemical composition of commercial citrus fruit essential oils and evaluation of their antimicrobial activity acting alone or in combined processes. *Food Control*, 22: 896-902.

European Union (EU)/SANCO. Scientific Committee on Veterinary Measures relating to Public Health (SCVPH) (2003). *Opinion of the SCVPH on Salmonellae in foodstuffs. Adopted on 14-15 April.* SCVPH, Brussels.

EUROSTAT data (2010). *Preparatory study on food waste across EU 27.* October 2010. European Commission (DG ENV). Available at: http://ec.europa.eu/environment/eussd/pdf/bio_foodwaste_report.pdf.

Fattouch, S., Caboni, P., Coroneo, V., Tuberoso, C. I. G., Angioni, A., Dessi, S., Marzouki, N., and Cabras, P. (2007). Antimicrobial Activity of Tunisian Quince (*Cydonia oblonga* Miller) Pulp and Peel Polyphenolic Extracts. *Journal of Agricultural and Food Chemistry*, 55, 963-969.

Ferrer, C., Ramón, D., Muguerza, B., Marco, A., and Martínez, A. (2009). Effect of Olive Powder on the Growth and Inhibition of *Bacillus cereus*. *Foodborne Pathogens and Disease,* 6, 1.

Forshell, L. P. and Wierup, M. (2006). *Salmonella* contamination: a significant challenge to the global marketing of animal food products. *Scientific and Technical Review of the Office International des Epizooties*, 25 (2), 541-554.

Ghafar, M. F. A., Prasad, K. N., Weng, K. K., and Ismail, A. (2010). Flavonoid, hesperidine, total phenolic contents and antioxidant activities from *Citrus* species. *African Journal of Biotechnology,* 9(3), 326-330.

Gracia, R. A. (2004). Evolución de la Industria Agroalimentaria Española en las dos últimas décadas. *Economía Industria*, 355(6): 197-200.

Govaris, A., Solomakos, N., Pexara, A., and Chatzopoulou, P. S. (2010). The antimicrobial effect of oregano essential oil, nisin and their combination against *Salmonella* Enteritidis in minced sheep meat during refrigerated storage. *International Journal of Food Microbiology*, 137, 175-180.

Gunduz, G. T., Gonul, S. A. and Karapinar, M. (2010). Efficacy of sumac and oregano in the inactivation of *Salmonella* Typhimurium on tomatoes. *International Journal of Food Microbiology*, 141, 39-44.

Igual, M., Sampedro, F., Martínez-Navarrete, N., and Fan, X. (2013). Combined osmodehydration and high pressure processing on the enzyme stability and antioxidant capacity of a grapefruit jam. *Journal of Food Engineering*, 114, 4, 514-521.

Karapinar, M. and Sengun, I. Y. (2007). Antimicrobial effect of koruk (unripe grape—*Vitis vinifera*) juice against *Salmonella* typhimurium on salad vegetables. *Food Control*, 18, 702-706.

Khan, M. K., Huma, Z. E. and Dangles, O. (2014). A comprehensive review on flavanones, the major citrus polyphenols. *Journal of Food Composition and Analysis*, 33, 86-104.

Kim, J., Marshall, M. R., Cornell, J. A., Preston, J. F., and Wei, C. I. (1995). Antibacterial activity of carvacrol, citral, and geraniol against *Salmonella* Typhimurium in culture medium and on fish cubes. *Journal of Food Science*, 60, 1364-1374.

Koutsoumanis, K., Lambropoulou, K. and Nychas, G. J. F. (1999). A predictive model for the non-thermal inactivation of *Salmonella* Enteritidis in a food model system supplemented with a natural antimicrobial. *International Journal of Food Microbiology*, 49, 63-74.

Losa, R. (2001). The use of essential oils in animal nutrition. In: Brufau, J. (ed.). *Feed manufacturing in the Mediterranean region. Improving safety: From feed to food.* Zaragoza: CIHEAM, 2001. p. 39-44 (Cahiers Options Méditerranéen nes; n. 54).

Mañas, P., Pagán, R., Raso, J., and Condón, S. (2003). Predicting thermal inactivation in media of different pH of *Salmonella* grown at different temperatures. *International Journal of Food Microbiology*, 87: 45-53.

Marín, F. R., Soler-Rivas, C., Benavente-García, O., Castillo, J., and Pérez-Álvarez, J. A. (2007). By-products from different citrus processes as a source of customized functional fibres. *Food Chemistry*, 100, 736-741.

Martín-Luengo, M. A., Yates, M., Diaz, M., Saez Rojo, E., and Gonzalez Gil, L. (2011). Renewable fine chemicals from rice and citric subproducts: Ecomaterials. *Applied catalysis b: Environmental*, 106, 488-493.

O'Shea, N., Arendt, E. K. and Gallagher, E. (2012). Dietary fibre and phytochemical characteristics of fruit and vegetable by-products and their recent applications as novel ingredients in food products. *Innovative Food Science and Emerging Technologies*, 16, 1-10.

Palaniappan, K. and Holley, R. A. (2010). Use of natural antimicrobials to increase antibiotic susceptibility of drug resistant bacteria. *International Journal of Food Microbiology*, 140, 164-168.

Puupponen-Pimia, R., Nohynek, L., Meier, C., Kahkonen, M., Heinonen, M., Hopia, A., and Oksman-Caldentey, K. M. (2001). Antimicrobial properties of phenolic compounds from berries. *Journal of Applied Microbiology*, 90, 494-507.

Sawalha, S. M. S., Arráez-Román, D., Segura-Carretero, A., and Fernández-Gutiérrez, A. (2009). Quantification of main phenolic compounds in sweet and bitter orange peel using CE-MS/MS. *Food Chemistry*, 116, 567-574.

Schieber, A., Stintzing, F. C. and Carle, R. (2001). By-products of plant food processing as a source of functional compounds – recent developments. *Trends in Food Science and Technology*, 12, 401-413.

Shapiro, R. S. and Cowen, L. E. (2012). Thermal Control of Microbial Development and Virulence: Molecular Mechanisms of Microbial Temperature Sensing. *American Society for Microbiology*, 3(5): 212-238.

Sobrino-López, A., Raybaudi-Massilia, R. and Martín-Belloso, O. (2006). High-intensity pulsed electric field variable affecting *Staphylococcus aureus* inoculated in milk. *American Dairy Science Association*, 89: 3739-3748.

Trípoli, E., La Guardia, M., Giammanco, S., Di Majo, D., and Giammanco, M. (2007). Citrus flavonoids: Molecular structure, biological activity and nutritional properties: A review. *Food Chemistry*, 204:466-479.

Valverde, M. T., Iniesta, F. M., Garrido, L. C., Rodrígez, A. T., García-García, I., Macanas, H., and Roda, R. C. (2010). Inactivation of *Salmonella spp* in refrigerated liquid egg products using essential oils and their active compounds. *Proceedings in International Conference on Food Innovation "Food Innova"*. Universidad Politécnica de Valencia.

Viuda-Martos, M., Ruiz-Navajas, Y., Fernández-López, J., and Pérez-Álvares, J. (2008). Antifungal activity of lemon (*Citrus Lemon* L.), mandarin (*Citrus reticulata* L.), grapefruit (*Citrus paradisi* L.) and orange (*Citrus sinensis* L.) essential oils. *Food Control,* 19: 1130-1138.

World Health Organisation (WHO). (1980). *Report of the WHO/World Association of Veterinary Food Hygienists (WAVFH) round table conference on the present status of the Salmonella problem (prevention and control)*, Bilthoven, the Netherlands, 6-10 October. WHO/VPH/81.27. WHO, Bilthoven.

Yang, S., Yu, R. and Chou, C. (2001). Influence of holding temperature on the growth and survival of *Salmonella* spp. and *Staphylococcus aureus* and the production of staphylococcal enterotoxin in egg products. *International Journal of Food Microbiology*, 63: 99-107.

In: Salmonella
Editor: Christopher B. Hackett
ISBN: 978-1-63463-651-3

Chapter 4

INVOLVEMENT IN FOODBORNE OUTBREAKS, RISK FACTORS AND OPTIONS TO CONTROL SALMONELLA ENTERITIDIS SE86: AN IMPORTANT FOOD PATHOGEN IN SOUTHERN BRAZIL

Eduardo Cesar Tondo,* Ana Carolina Ritter and Letícia Sopeña Casarin
Instituto de Ciência e Tecnologia de Alimentos, Universidade Federal do Rio Grande do Sul – ICTA/UFRGS, Brazil

ABSTRACT

A specific strain of *S.* Enteritidis (SE86) was responsible for more than 95 % of the investigated salmonellosis occurred in the State of Rio Grande do Sul (RS), Southern Brazil, from 1999 to 2013. The aim of this chapter is to demonstrate the involvement of this strain with the salmonellosis outbreaks and to discuss the factors that probably contributed to *S.* Enteritidis SE86 had become one of the most important food pathogens in Southern Brazil. During 1999 to 2006, 190 salmonellosis outbreaks were investigated in RS and one DNA banding profile was identified among *S.* Enteritidis isolated from foods, blood and feces of victims. The causative strain was named *S.* Enteritidis SE86. The main risk factors for the outbreaks caused by SE86 were 1) the consumption of homemade mayonnaise prepared with raw eggs; 2) holding foods in room temperature for more than 2 hours, 3) cross-contamination due to the contact of food with contaminated equipment and utensils. The growth of SE86 in homemade mayonnaise was modeled and results demonstrated that SE86 was able to grow faster than others *Salmonella* serovars, during the first six hours at environmental temperature; however SE86 did not grow in homemade mayonnaise at 10° C, during 18 hours. Further studies have demonstrated that SE86 was able to form biofilms on stainless steel, stainless steel welds and polypropylene surfaces, and survived to 400 and 200 ppm sodium hypochlorite. After sodium hypochlorite exposure, SE86 expressed RpoS and Dps proteins, which are involved with oxidative stress. Due to exposure to sub-lethal pH, SE86 became acid-adapted and

* Corresponding author: tondo@ufrgs.br.

increased its thermal resistance. Acid-adapted SE86 was able to survive to simulated gastric fluid pH 1.5 and became more virulent than other *Salmonella* serovars for germ-free mice. Isolates of SE86 remain sensitive to several antibiotics, nevertheless resistance to ampicillin and nalidixic acid and multidrug resistance has increased during last decade.

INTRODUCTION

Foodborne Diseases in Southern Brazil

Currently, Brazil exports foods to approximately 160 countries and is a major exporter of poultry, beef, coffee, sugarcane, soybeans, corn, pork, and cocoa [1]. In order to attend the high standards demanded by exportation, high quality and safe foods are produced by Brazilian food industries. However, in a country with approximately 200 million people, several different realities can be found, including in food production. Similar to other countries, many establishments prepare foods without adequate conditions, resulting in food-borne outbreaks that, in most of the cases, are not notified or registered. The majority of the food-borne outbreak notifications came from the Southern and Southeast regions of Brazil, because the surveillance services are better structured in these regions. The occurrence of food-borne outbreaks in other regions is not clear [2]. The Southeast region of Brazil is composed by the States of Espírito Santo, Rio de Janeiro, Minas Gerais and São Paulo, while the Southern region grouped the States called Paraná, Santa Catarina, and Rio Grande do Sul. These regions are well developed, presents several industrialized areas and high standards of life quality. Rio Grande do Sul (RS) is the southernmost State of Brazil and has a population of 10.7-million-people, distributed in 496 cities [3].

In Brazil, *Salmonella* sp. was identified as the major cause of reported food-borne diseases, during the period from 2000 to 2013 [4]. Among the 3,857 etiologic agents identified in food-borne illnesses, *Salmonella* sp. was identified in 39.46 % of the outbreaks, followed by *Staphylococcus aureus* (19.68 %), *Escherichia coli* (12.32 %), and *Bacillus cereus* (7.65 %) [4]. Unfortunately in approximately 50 % of the investigated food-borne outbreaks, the etiological agent was not identified [5, 6]. Considering the whole country, mixed food preparations and eggs and egg-based products, respectively, were the most common food vehicles of food-borne disease outbreaks. According to the Brazilian official data, the majority of outbreaks occurred in private homes, followed by restaurants, bakeries and schools [4].

Salmonellosis in Southern Brazil

During the 1990s, *Salmonella* spp. became the major cause of reported food-borne diseases in the State of RS [7], surpassing *S. aureus*, which was identified as the major agent of food-borne diseases of this State, from 1987 to 1993. According to the Sanitary Surveillance Service of RS (DVS/RS), the reason for that was the implementation of mandatory Good Manufacturing Practices (GMP) training courses for people responsible for food services of RS. The better control of food manipulation decreased the contamination originated by food handlers, reducing outbreaks caused by *S. aureus* [3].

Official data of DVS/RS about food-borne salmonellosis of RS were analyzed by periods (1997 to 1999, 2001 to 2002, and 2003 to 2004) and published elsewhere [8, 9, 10]. Data have demonstrated that more than 46,000 people were involved, and approximately 5,200 people were hospitalized, during all the periods. Most of salmonellosis outbreaks occurred in springtime, due to keep foods outside refrigeration. This inadequate procedure occurred because, at springtime, RS presents mild temperatures at mornings and people conclude that refrigeration of foods is not necessary. At afternoons, the temperature increases and microorganisms multiply on foods, causing the outbreaks. The incidence of outbreaks caused by other etiological agents also increases, confirming the negligence regarding control of temperature [8]. At summertime, the number of salmonellosis usually decreases because population is aware about food-borne diseases, storing foods in adequate refrigeration.

From 1997 to 2004, in RS, the most frequent food vehicle of *Salmonella* was potato salad mixed with homemade mayonnaise prepared with raw eggs [8, 9, 10]. A recent study reported that this preparation remained as the most important food vehicle of salmonellosis in RS [11], even though other food preparations were also identified, for example, pastry products (15.94 %), beef (12.32 %), processed meat (9.42 %), chicken meat (6.52 %), and pork (2.17 %) [11]. According to Vaz, et al. (2010) [12], in Brazil and in other parts of the world, *S.* Enteritidis has been associated with human foodborne infections caused by the ingestion of contaminated foods of animal origin, mainly undercooked poultry meat and eggs. Nunes, et al. (2003) [13], analyzed *Salmonella* isolated in 1995 to 1997 from Brazilian healthy and diseased chicken, food-borne outbreaks of human gastroenteritis related to the consumption of egg products, poultry meat, pipped embryos of broiler chicks, meat meals, and diverse food products, as cheese, mayonnaise, cake, and bacon and reported that phage type (PT) 4 was the most frequent type of *S.* Enteritidis found. The same authors stated that the results obtained were in accordance with the worldwide trends in distribution patterns for different *S.* Enteritidis phage types [13]. Pires, et al. (2014) [14] studied the source attribution of human salmonellosis worldwide and demonstrated illnesses and outbreaks were most commonly attributed to exposure to contaminated food, mainly eggs, broiler chickens, and pigs. The same authors also mentioned that exposure to raw vegetables was also an important source of salmonellosis.

Salmonella serovars responsible for food-borne outbreaks in RS have changed along the years. For example, in 1990, *S.* Typhimurium was the predominant *Salmonella* serovar (86 %) isolated from foods involved with salmonellosis of RS, while *S.* Enteritidis was not found in those samples [8]. However, in 1993, *S.* Enteritidis was isolated from 64 % of the suspected foods, surpassing *S.* Typhimurium, which was isolated from only 4 % of the food involved with salmonellosis [8]. During the 90s, this surpassing of *S.* Typhimurium by *S.* Enteritidis was also observed in the United States and other countries [15]. The rapid increase in the prevalence of *S.* Enteritidis in different countries was attributed to the formation of an ecological niche occurred due to the successful campaign to eradicate *S.* Gallinarum and *S.* Pullorum from chicken flocks, which are the causative agents of fowl typhoid and pullorum disease in chickens [16, 17].

Geimba, et al. (2004) [16] analyzed 75 *Salmonella* spp. isolated from foods involved with food-borne outbreaks occurred from 1999 to 2000 in RS and found that 73 (97 %) of them were *S.* Enteritidis. Only one isolate was serotyped as *S.* Derby and other one as *S.* Typhimurium. Among the strains of *S.* Enteritidis, 83 % were positive for the presence of *spv*R gene (*Salmonella* Plasmid Virulence regulatory gene) and a positive correlation (*P*

<0.05) between *S.* Enteritidis and the presence of *spv*R was reported [16, 3]. In the period from 2001 to 2002, Oliveira, et al. (2007) [18] investigated 85 *Salmonella* isolated from foods responsible for salmonellosis occurred in RS and observed that 93 % were *S.* Enteritidis. Only six isolates were serotyped as *S.* Javiana (1), *S.* Infantis (1), *S.* Agona (1), *S.* Typhimurium (1), and *S. enterica* subspecies enterica 1, 4 [18].

Recently, Capalonga et al. (2014) [11] reported that 84.7 % of the 163 *Salmonella* isolated from foods involved with salmonellosis in RS, during 2007 to 2012, were *S.* Enteritidis. The second and third most frequent serovars were *S.* Schwarzengrund (5.5 %) and *S.* Typhimurium (3.7 %).

Molecular Characterization of *S.* Enteritidis in Southern Brazil

Oliveira, et al. (2007) [18] demonstrated that *S.* Enteritidis responsible for salmonellosis in RS during the years of 2001-2002 were grouped in only two PCR-ribotyping banding profiles (R1 and R2), being that profile R1 comprised 92.4 % of the strains. The same strains were analyzed by RAPD and demonstrated four banding patterns (A to D). Profile A grouped 81 % of the strains. Using PCR ribotyping, Pulsed Field Gel Electrophoresis (PFGE) and DNA sequencing, Oliveira, et al. (2009) [19] investigated the clonal relationship of 152 isolates of *S.* Enteritidis involved with food-borne outbreaks of RS, from 1999 to 2002, concluding that a specific strain of *S.* Enteritidis was found in the majority of the investigated outbreaks. After that, during 1999 to 2006, the Official Laboratory of RS (FEPPS/LACEN/RS) isolated 931 *S.* Enteritidis isolates in 190 outbreaks officially investigated. Among them, *S.* Enteritidis isolates from 80 different outbreaks were randomly chosen to be analyzed (the researchers picked ten isolates from each year of the period comprised from 1999 to 2006). The 80 *S.* Enteritidis samples were isolated from blood (n=12) and feces (n=68) of salmonellosis victims in RS, and were analyzed by means of PCR ribotyping and XbaI macro restriction PFGE. Results identified only three closely related PCR ribotyping patterns (R1, R2, and R3) and just one PFGE profile. The PCR ribotyping profile were the same profile previously found in *S.* Enteritidis isolated in suspected foods [18]. The PFGE profile grouped 97 % of the strains isolated from 1999 to 2006 [20]. These findings confirmed the occurrence of the same DNA banding pattern among *S.* Enteritidis isolated in foods, blood and feces of the victims of salmonellosis from 1999 to 2006 in RS, suggesting that a specific strain of *S.* Enteritidis was responsible for the majority of the cases of salmonellosis in this Brazilian State. This strain was named *S.* Enteritidis SE86 [3].

Lately, Capalonga, et al. (2014) [21] studied 163 *S.* Enteritidis isolated from foods involved in foodborne salmonellosis occurred in RS from 2007 to 2013 and found only one PCR ribotyping profile, which was compatible to the profile of SE86. Based on these results, the researchers concluded that SE86 have caused several salmonellosis outbreaks in RS, from 1999 to 2013.

Risk Factors for Salmonellosis in Southern Brazil

According to DVS/RS [8, 9, 10, 3], the risk factors for salmonellosis outbreaks in RS were: 1) the consumption of homemade mayonnaise prepared with raw eggs (use of eggs without sanitary inspection), 2) holding foods in room temperature for more than 2 hours, 3) cross-contamination due to the contact of foods with surfaces of contaminated equipment and utensils, and 4) maintenance of food in improper cooling. These risk factors, as well as studies related to them are discussed below.

Consumption of Homemade Mayonnaise

Homemade mayonnaise (HM) was identified as the food vehicle mostly involved with salmonellosis in Southern Brazil [22]. It is well known that eggs and egg products are among the most important food vehicles of *S.* Enteritidis, because the outer shell egg surfaces or the internal egg contents can be contaminated [23]. In general, the external contamination is due to chicken feces during or after oviposition, on the other hand the internal infection can be the result of penetration through the eggshell or by direct contamination of egg contents, before oviposition, originating from reproductive organs infection [24]. Mayonnaise, often prepared with raw eggs, is widely consumed and probably the most used sauce around the world [25]. It is a food preparation frequently involved with foodborne outbreaks worldwide [26, 27, 28, 29]. In the State of RS, HM prepared with raw eggs and mixed with cooked potato is the principal side dish of *churrasco*, a typical meat meal similar to barbecue, highly consumed in this region of Brazil [22, 30].

In order to allow the development of a quantitative microbial risk assessment for *S.* Enteritidis SE86 on HM, a study was performed in order to modeling the growth of SE86 on HM at different temperatures [22]. Using predictive microbiology programs, growing curves were built by fitting data to the Baranyi's DMFit, and generated R^2 values greater than 0.98 for primary models. Secondary model was fitted with Ratkowsky equation, generating R^2 and RMSE values of 0.99 and 0.016, respectively. Experimental data showed that SE86 did not grow on HM at 7°C, for 30 days. At 10°C, no growth was observed until approximately 18 h and the infective dose (considered 10^6 CFU) was reached after 8.1 days. However, 10^6 CFU of SE86 were attained after 6 h at 37°C. Experimental data demonstrated smaller *lag* phases than those generated by ComBase Predictive Models, suggesting that SE86 is very well adapted for growing on HM. Corroborating this finding, Malheiros, et al. (2007) [31] demonstrated that SE86 was able to grow faster than others *Salmonella* serovars (Bredeney and Typhimurium) during the first six hours on HM stored at environmental temperature. The frequent association of *S.* Enteritidis with eggs and this faster growing can be factors that contribute to the frequent involvement of SE86 with salmonellosis caused by the consumption of HM. Elias, et al. (2014) [22] reported that a safe HM can be prepared if non-contaminated eggs (or cooked yolks) are used and HM are stored below 7° C.

Cross-Contamination by Contaminated Equipment and Utensils

According to DVS/RS, cross-contamination was the third most important risk factor for salmonellosis in RS [8, 9, 10, 3]. Based on this, the biofilm formation capability of SE86 was investigated on different material surfaces. For example, Tondo, et al. (2010) [32] immersed coupons of stainless steel and polyethylene in bacterial suspensions of SE86, *S.* Typhimurium, and *S.* Bredeney, during 15, 30, and 60 minutes. The three serovars showed similar counts of adhered cell to both materials (5.0 to 6.5 log CFU. cm^{-2}), and the time of exposure did not influence the counts of adhered cells on both surfaces; however, Scanning Electron Microscopy revealed larger clusters of SE86 on both materials, which was not found for the other serovars [32].

Casarin, et al. (2014) [33] investigated the adhesion of *S.* Enteritidis SE86 and *Listeria monocytogenes* on stainless steels (types 304 and 316) after 0, 1, 2, 4, 6, and 8 h of contact time, at room temperature. The study also evaluated the influence of the material topography and the hydrophobicity of cells and material surfaces in the attachment process. Results demonstrated that both bacteria were able to adhere on both types of stainless steels. However, at the beginning of contact time between cells and surfaces, higher numbers of adhered cells of *S.* Enteritidis SE86 were observed. S. Enteritidis SE86 also demonstrated a higher negative total energy of adhesion than *L. monocytogenes*, meaning that the adhesion of SE86 was thermodynamically more favorable than the adhesion of *L. monocytogenes* to both types of stainless steels [33]. The hydrophobicity of the investigated surfaces was also determined based on contact angle measurements. According to the results, both bacteria were hydrophilic and stainless steel surfaces were hydrophobic. Despite of the different levels of hydrophobicity/hydrophilicity, there was no correlation between adhesion and the surface hydrophobicity [33].

In another study Casarin, et al. (2014) [34] investigated the adhesion of *S.* Enteritidis SE86 and *L. monocytogenes* on the surface of metal inert gas (MIG), and tungsten inert gas (TIG) welds. Results demonstrated that *S.* Enteritidis SE86 initially adhered significantly more than *L. monocytogenes* on both types of welds, and there was a significant difference ($P<0.05$) in the adhesion of SE86 when 0h and 1h of contact time were compared. According to the data obtained, bacteria were hydrophilic, while weld surfaces were hydrophobic [34].

Resistance to Differents Sanitizers

Machado, et al. (2010) [35] evaluated the resistance of *S.* Enteritidis SE86, *S.* Typhimurium, and *S.* Bredeney to sanitizers, using the suspension test recommended by the Brazilian Ministry of Agriculture. Three commonly used sanitizers in food industries were tested: peracetic acid, quaternary ammonium and sodium hypochlorite. The results demonstrated that SE86 was resistant to 400 ppm sodium hypochlorite and survived for up to 15 minutes of exposure to 200 ppm of this sanitizer, a fact which was not demonstrated by the other strains [35]. In other study, Tondo, et al. (2010) [32] tested the same microorganisms to biofilm formation capability on surfaces of material commonly used in food industries and food services. Coupons of stainless steel and polyethylene were immersed in cultures of SE86, *S.* Typhimurium, and *S.* Bredeney during 15, 30 and 60 minutes. After that, coupons were exposed to different concentrations of peracetic acid, sodium hypochlorite, and quaternary ammonium [32]. The sanitizers did not inactivate all the microorganisms adhered on both materials, being that least 1 log CFU. cm^{-2} of the microorganisms remained viable [32]. The study demonstrated that *S.* Typhimurium and SE86 were more resistant to quaternary ammonium than *S.* Bredeney on polyethylene surface, and the reduction of SE86 was smaller than the other serovars after sodium hypochlorite treatments [32].

The resistance to sodium hypochlorite demonstrated by SE86 was compared to the resistance of other pathogenic strains of *S.* Enteritidis isolated from Albania, Zimbabwe,

Morocco, and Pakistan [32, 35]. The strains were exposed to 200 ppm sodium hypochlorite for 5, 10, 15, and 20 minutes, according recommendations of the Brazilian Ministry of Agriculture. Results demonstrated that none of the *S*. Enteritidis were totally inactivated after 20 minutes of exposure and the reduction rates were similar. Only the strain isolated in Albania demonstrated to be significantly more sensitive [36].

The rule of *rpo*S and *dps* genes in the resistance to 200 ppm sodium hypochlorite was investigated in SE86. Mutants of SE86 were constructed using the method described by Uzzau, et al. (2001) [37]. The survival of the Wild Type (WT) strain, as well as of the attenuated strains, was determined by bacterial counts. Tagged proteins (Dps and RpoS) were detected by means of SDS-PAGE and also immunoblotting with anti-FLAG antibodies. SE86 lacking *dps* demonstrated greater sensitivity compared to the WT SE86 exposed to sodium hypochlorite. The RpoS and Dps proteins were actively expressed under the conditions investigated, suggesting that these SE86 genes are related to oxidative stress caused by sodium hypochlorite [36].

Acid and Thermal Resistance of *S*. Enteritidis SE86

Malheiros, et al. (2008) [38] cultivated SE86, *S*. Typhimurium, and *S*. Bredeney in Nutrient Broth (NB), as well as in Nutrient Broth supplemented with 1 % glucose (NBG). The latter was used in order to induce acid adaptation of cells exposed to sub-lethal pH generated by the degradation of glucose. The acid-adapted and non acid-adapted microorganisms were exposed to different pH (3.5, 4.0, and 4.5) and temperatures (52, 56, and 60 ºC), and survival curves were created [38]. Results showed that *S*. Bredeney demonstrated higher resistance to pH 3.5 and 4.0; nevertheless, SE86 demonstrated a better capacity for acid adaptation (7.5 times greater) than other *Salmonella* serovars. At pH 4.5, all serovars demonstrated a similar acid resistance, remaining at the same levels of viable cells for 300 minutes. At 52 ºC, acid adaptation was able to protect only SE86. At 56 ºC and 60 ºC, non-adapted and acid-adapted SE86 strains were more thermally resistant than other serovars tested, demonstrating that SE86 was more thermal resistant than other strains. SDS-PAGE analysis demonstrated differences in the protein profiles of non-adapted and acid-adapted cells of all serovars [38].

Perez, et al. (2012) [39] evaluated the survival and the capability of intestinal invasion of SE86 and *S*. Typhimurium (ST99), after acid adaptation. Both strains were cultivated in Nutrient Broth supplemented with 1 % glucose (NBG) and were exposed to simulated gastric fluid (FGS) pH 1.5. In a second step, approximately 8 log of both strains (either acid-adapted or non-adapted) were orally inoculated in germ-free adult male *Wistar* mice. Animals were observed at aseptic conditions for twelve days [39]. Animal feces and portions of the gastrointestinal tract were analyzed microbiologically, and the appearance of intestinal morphological abnormalities was investigated [39]. Animals were submitted to the mortality curve. The results indicated that acid-adapted SE86 had a significant higher survival rate ($p < 0.05$) than non-adapted SE86, non-adapted ST99 and also than acid-adapted ST99 after exposure to FGS. The *in vivo* experiments demonstrated that acid-adapted SE86 and acid-adapted ST99 were able to cause intestinal morphological abnormalities [39]. Acid-adapted SE86 showed higher counts in the ileum-cecal junction than the other strains, suggesting that acid adaptation influenced the virulence of this strain [39, 3]. All strains were able to rapidly multiply in germ-free mice; however mortality caused by acid-adapted SE86 was more intense [39]. Histopathological analyses revealed greater severity of the infection caused by SE86, a fact which was confirmed by the death of animals starting at the fourth day of infection. ST99 did not cause the death of animals until twelve days after inoculation [39].

The ability of *Lactobacillus acidophilus* LA10 to colonize and exert antagonistic effects against SE86 in the gastrointestinal tract of conventional mice was analyzed [40]. Doses of 10^8 viable cells of SE86 and *L. acidophilus* LA10 were administered by gavage to mice. The experiment used 4 groups of mice. Group 1 was administered only sterile saline solution and was considered the negative control. Group 2 was administered only SE86. Group 3 was first administered SE86 and, after 10 days, treated with *L. acidophilus* LA10. Group 4 was first administered *L. acidophilus* LA10 and, after 10 days, challenged with SE86. The results demonstrated that a significant number of SE86 cells were able to colonize the gastrointestinal tract of mice, specifically in the colon and ileum. *L. acidophilus* LA10 demonstrated antagonistic effect against SE86, with better results observed for Group 3 over Group 4. Thus, *L. acidophilus* LA10 demonstrated potential antagonistic effects against *S.* Enteritidis SE86, especially if administered after infection [40].

Antimicrobial Resistance of *S.* Enteritidis SE86

Even though the salmonellosis caused by *S.* Enteritidis usually is limited to gastrointestinal tract and the treatment does not involve antibiotics, the use of these drugs is recommended when salmonellosis affects immunocompromised patients or when the symptoms of salmonellosis are more severe such as the presence of blood in stools and fever [11]. The concern is that if resistant strains cause food poisoning in these patients they can be difficult to treat, making the probability of human death more likely. In the last decades, the resistance of *Salmonella* sp. to antibiotics has increased rapidly throughout the world, mainly due to indiscriminate and incorrect use of antibiotics [41, 3]. In spite of *S.* Enteritidis strains present low antimicrobial resistance when compared with some isolates of *S.* Typhimurium [42], attention should be given to the frequent isolation of multidrug-resistant *S.* Enteritidis [43]. Examples of that were reported in Tondo and Ritter (2012) [3].

The antibiotic resistance of *S.* Enteritidis isolated from foods involved with salmonellosis in the State of RS State has been investigated by several studies. Most of the isolates demonstrated to be sensitive for the majority of the drugs evaluated, however resistance to streptomycin, gentamicin, and nalidixic acid were observed among microorganisms isolated in 1999 to 2000 and in 2001 to 2002. Few isolates exhibited multidrug resistance [3]. Oliveira, et al. (2012) [44] studied 80 *S.* Enteritidis isolated from salmonellosis in RS from 1999 to 2006, and observed that only three isolates (4 %) were fully susceptible to all the antibiotics tested. The major resistances were observed for ampicillin (81 %), streptomycin (19 %), and nalidixic acid (25%) [44, 3]. In the period of 2003 to 2006, Paula, et al. (2011) [45] evaluated 130 *S.* Enteritidis isolates responsible for salmonellosis occurred in RS and reported that the higher percentages of resistance were demonstrated for ampicillin (100 %) and nalidixic acid (48 %). When compared to previous studies [44, 46], results demonstrated that multi-resistance increased expressively, because 63 % of the isolates showed multi-resistance [45].

Among the 138 *S.* Enteritidis isolated from salmonellosis in the period from 2007 to 2012, the highest resistance percentages observed were for nitrofurantoin (94.2 %) and nalidixic acid (89.1 %). Only two isolates were resistant to tetracycline, two to ampicillin, and two to cefoxitin. Three isolates were resistant to ceftazidime, ciprofloxacin, and trimethoprim/sulfamethoxazole, separately, while no isolate showed resistance to chloramphenicol, streptomycin, imipenem, and gentamicin [11]. Comparing the results of Geimba, et al. (2005) [46], Oliveira, et al. (2012) [44], De Paula, et al. (2011) [45], and Capalonga, et al. (2014) [11], it is possible to observe that the percentage of *S.* Enteritidis resistant to ampicillin and nalidixic acid is increasing over the years. For example, Geimba, et al. (2005) [46] observed that in 1999 12.8 % of the *S.* Enteritidis isolates were resistant to nalidixic acid, and in 2000 this percentage increased to 14.7 %. Oliveira, et al. (2012) [44] showed an increase in the resistance to the same antibiotic from 19.0 % to 24.3 %, between strains isolated from 2001 and 2002. De Paula, et al. (2011) [45] showed an increase from 40.7 % to 66.7 %, during the period of 2003 to 2006. Finally, Capalonga, et al. (2014) [11] demonstrated that, in 2007, 89.1 % of the isolates were resistant to nalidixic acid, increasing to 100 %, in 2012.

The raising resistance for nalidixic acid is issue of concern, because several studies have reported increasing numbers of *Salmonella* sp. resistant to quinolones in Germany [48], England and Wales [49], and Spain [50]. Emphasizing this concern, the CLSI added a guideline recommending clinical laboratories to routinely test for nalidixic acid resistance in extra intestinal *Salmonella* isolates in order to alert physicians about the emerging resistance [51]. Quinolones are widely used in food animal production and are able to select quinolone-resistant *Salmonella* sp. in animals. At the same time, quinolones are one of the few available therapies for serious *Salmonella* infections, particularly in adults.

Conclusion

S. Enteritidis SE86 has been involved with several food-borne salmonellosis occurred in the State of RS, during 1999 to 2013. The most important risk factors for salmonellosis caused by this strain were the consumption of homemade mayonnaise prepared with raw eggs, holding foods in room temperature for more than 2 hours; and cross-contamination due to the contact of foods with surfaces of contaminated equipment and utensils. SE86 was able to form biofilms on stainless steel AISI 316 and AISI 304, polyethylene, MIG and TIG weld surfaces, and survived to 400 and 200 ppm sodium hypochlorite. After being exposed to sub-lethal pH, SE86 became acid-adapted. The acid-adapted SE86 became more resistant to acid and thermal exposures and was more virulent to mice. In homemade mayonnaise, SE86 has grown faster than others *Salmonella* serovars (Bredeney and Typhimurium), during the first six hours at environmental temperature, however was not able to grow at 7° C. During the last decade, SE86 has shown increasing rates of resistance to ampicillin and nalidixic acid, but still sensitive to antibiotics like chloramphenicol, streptomycin, imipenem, and gentamicin.

ACKNOWLEDGMENTS

We would like to thank all the Professionals of Food Sector of DVS/CEVS/RS and from Water and Food Microbiology Laboratory of FEPPS/LACEN/RS who have worked intensively to improve the regional Surveillance Service of RS.

REFERENCES

[1] Brazil. Ministério da Agricultura, Pecuária e Abastecimento. (2009). Available at: http://www.agricultura.gov.br, accessed: January 14, 2012.

[2] Gomes, B.C., Franco, B.D.G.M., Martinis, E.C.P. (2013). Microbiological food safety issues in Brazil: bacterial pathogens. *Foodborne Pathogens and Disease*, 10, 3, 197-205.

[3] Tondo, E.C., Ritter, A.C. (2012). *Salmonella* and Salmonellosis in Southern Brazil: a review of the last decade - Salmonella: Classification, Genetics and disease Outbreaks, Nova Science Publishers, Inc, Nova York, USA.

[4] Brazil. Ministério da Saúde. Secretaria de Vigilância em Saúde - SVS (2013). Dados Epidemiológicos- DTA – Período de 2000 a 2013. Available at: http://www.saude.gov.br/svs, accessed: June 30, 2014.

[5] Brazil. Ministério da Saúde. Secretaria de Vigilância em Saúde - Vigilância Epidemiológica do Botulismo. (2008). Available at: http://portal.saude.gov.br/ portal/ arquivos/pdf/Botulismo.ppt.pdf, accessed: January 12, 2012.

[6] Brazil. Ministério da Saúde. Secretaria de vigilância em Saúde. Vigilância epidemiológica das doenças de transmissão hídrica e alimentar - VEDTHA. (2011). Available at: http://portal.saude.gov.br/portal/arquivos/pdf/10_passos_para_ investi gacao_surtos.pdf, accessed: June 7, 2012.

[7] Rio Grande do Sul. Secretaria Estadual da Saúde. (2001) Divisão de Vigilância Sanitária. Relatórios anuais de DTA. Porto Alegre.

[8] Costalunga, S. and Tondo, E. C. (2002) Salmonellosis in Rio Grande do Sul, 1997 to 1999. *Brazilian Journal of Microbiology*, 33, 342-346.

[9] Silveira, J.B. and Tondo, E.C. (2006) Salmonellosis outbreaks occurring in Rio Grande do Sul, Southern Brazil, during 2000 to 2001. In: *International Symposium Salmonella and Salmonellosis* (I3S), Saint-Malo. Proceedings. v. 1. p. 521-522.

[10] Wagner, V.R., Silveira, J.B., Tondo, E.C. (2013). Salmonellosis in the State of Rio Grande do Sul, southern Brazil, 2002 to 2004. *Brazilian Journal of Microbiology*, 44(3), 723-729.

[11] Capalonga, R., Ramos, R.C., Both, J.M.C., Soeiro, M.L.T., Longaray, S.M., Simone Haas, S., Tondo, E.C. (2014) *Salmonella* serotypes, resistance patterns, and food vehicles of salmonellosis in southern Brazil between 2007 and 2012. *Journal of Infections in Developments Countries*, 8 (7), 811-817.

[12] Vaz, C.S., Streck, A.F., Michael, G.B., Marks, F.S., Rodrigues, D.P., Dos Reis, E.M., Cardoso, M.R., Canal, C.W. (2010) Antimicrobial resistance and subtyping of Salmonella enterica subspecies enterica serovar Enteritidis isolated from human outbreaks and poultry in southern Brazil. *Poultry Science*, 89, 1530–1536.

[13] Nunes, I.A., Helmuth, R., Schroeter, A., Mead, G.C., Santos, M.A.A., Solari, C.A, Silva, O.R., Ferreira, A.J. (2003). Phage typing of Salmonella enteritidis from different sources in Brazil. *Journal of Food Protection*, 66, 324–327.

[14] Pires, S.M., Vieira, A.R., Hald, T., Cole, D. (2014) Source Attribution of Human Salmonellosis: An Overview of Methods and Estimates. *Foodborne Pathogens and Disease*, 11 (9), 667-676.

[15] Usera, M.A., Popovic, T., Bopp, C., Strockbine, N.A. (1994) Molecular subtyping of *Salmonella enteritidis* phage type 8 strains from United States. *Journal of Clinical Microbiology*, 32, 194-198.

[16] Geimba, M.P., Tondo, E.C., Oliveira, F.A., Cardoso, M.I. (2004) Serological characterization and prevalence of spvR genes in *Salmonella* isolated from foods involved in outbreaks in *Brazil. Journal of Food Protection*, 67, 1229-1233.

[17] Bäumler, A.J., Hargis, B.M., Tsolis, R.M. (2000) Tracing the origins of *Salmonella* outbreaks. *Science*, 287, 50-52.

[18] Oliveira, F.A., Frazzon, A.P., Brandelli, A., Tondo, E.C. (2007) Use of PCR-ribotyping, RAPD, and antimicrobial resistance for typing of *Salmonella* Enteritidis involved in food-borne outbreaks in Southern *Brazil. Journal of Infections in Developments Countries*, 1, 170-176.

[19] Oliveira, F.A., Geimba, M.P., Pasqualotto, A.P. and Tondo, E.C. (2009) Clonal relationship among *Salmonella* enterica serovar Enteritidis involved. *Food Control*, 20, 606-610.

[20] Oliveira, F.A., Pasqualotto, A.P., Padilha, W., Tondo, E.C. (2010) Characterization of *Salmonella* Enteritidis isolated from human samples. *Food Research International*, 1, 2-5.

[21] Capalonga, R., Ramos, R; C., Both, J.M.C., Soeiro, M.L.T., Solange Mendes Longaray, S.M, Haas, S, Tondo, E.C. (2014). Characterization of Salmonella Enteritidis involved in foodborne outbreaks in Southern Brazil, from 2007 to 2013. *Journal of Infections in Developments Countries* (submitted).

[22] Elias, S.O., Sant'Ana, A.S., Tondo, E.C. Modeling growth kinetic parameters of *Salmonella* Enteritidis SE86 on Homemade Mayonnaise. *Food Microbiology* (submitted).

[23] Zhang, G., Brown, E.W., Hammack, T.S. (2013). Comparison of different preenrichment broths, egg: preenrichment broth ratios, and surface disinfection for the detection of *Salmonella* enterica ssp. enterica serovar Enteritidis in shell eggs. *Poultry Science*, 92, 3010-3016.

[24] Gantois, I., Ducatelle, R., Pasmans, F., Haesebrouck, F., Gast, R., Humphrey, T.J., Van Immerseel, F. (2009). Mechanisms of egg contamination by Salmonella Enteritidis. *FEMS Microbiology*, 33, 718–738.

[25] Guerra, C.B. and Miguel, D.P. (2010). Positive staphylococcus coagulase and fecal coliforms in cold salads added to mayonnaise sauce. *Food Research International*, 1, 11-15.

[26] Fica, A., Acosta, G., Dabanch, J., Perret C., Torres, M., López, J., Jofré, L., Weitzel, T. (2012). Salmonellosis outbreaks and the size and role of the Chilean State. *Revista Chilena de Infectología*, 29, 207-214.

[27] Norton, S., Huhtinen, E., Conaty, S., Hope, K., Campbell, B., Tegel, M., Boyd, R., Cullen, B. (2012). A large point-source outbreak of *Salmonella* Typhimurium linked to

chicken, pork and salad rolls from a Vietnamese bakery in Sydney. *Western Pacific Surveillance and Response Journal*, 21, 16-23.

[28] Di Pinto, A., Novello, L., Montemurro, F., Bonerba, E., Tantillo, G. (2010). Occurrence of Listeria monocytogenes in ready-to-eat foods from supermarkets in Southern Italy. *New Microbiologica*, 33, 249-252.

[29] Stein-Zamir, C., Tallen-Gozani, E., Abramson, N., Shoob, H., Yishai, R., Agmon, V., Reisfeld, A., Valinsky, L., Marva, E. (2009). *Salmonella* enterica outbreak in a banqueting hall in Jerusalem: the unseen hand of the epidemiological triangle. *Israel Medical Association Journal*, 11, 94-97.

[30] Albrecht, C.F. (2010) Além da carne assada sobre brasas: os elementos da experiência de consumo do churrasco (dissertation) UFRGS, Porto Alegre, Brazil.

[31] Malheiros, P. S., Paula, C. D. M. and Tondo, E. C. (2007) Avaliação da cinética de crescimento de *Salmonella* Enteritidis envolvida em surtos alimentares no RS: uma comparação com outros sorovares. *Ciência e Tecnologia de Alimentos*, 27, 751-755.

[32] Tondo, E.C., Machado, T.R.M., Malheiros, P.S., Padrão, D.K., Carvalho, A.L., Brandelli, A. (2010). Adhesion and biocides inactivation of *Salmonella* on stainless steel and polyethylene. *Brazilian Journal of Microbiology*, 4, 10-20.

[33] Casarin, L.C., Casarin, F.O., Brandelli, A., Novello, J., Ferreira, S.O., Tondo, E.C. (2014). Attachment of *Salmonella* Enteritidis and Listeria monocytogenes to stainless steels AISI 304 and AISI 316: the free energy influence. *Brazilian Journal of Microbiology (submitted).*

[34] Casarin, L. S., Brandelli, A. Fabrício de Oliveira Casarin, F. O. Paulo, A. S., Wanke, C. H., Tondo, E. C. (2014). Adhesion of *Salmonella* Enteritidis and *Listeria monocytogenes* on stainless steel welds. *International Journal of Food Microbiology*, 191, 103–108.

[35] Machado, T.R.M., Malheiros, P.S., Brandelli, A., Tondo, E.C. (2010) Avaliação da Resistência de *Salmonella* à ação de desinfetantes ácido peracético, quaternário de amônio e hipoclorito de sódio. *Revista do Instituto Adolfo Lutz*, 69, 475-481.

[36] Ritter, A.C., Bacciu, D., Santi, L. Silva, W.O.B., Vainstein, M. H., Rubino, S., Uzzau, S. Tondo, E. C. (2011) Investigation of *rpoS* and *dps* genes in sodium hypochlorite resistance of *Salmonella* Enteritidis SE86 isolated from foodborne illness outbreaks in Southern Brazil. *Journal of Food Protection*, 75, 437-442.

[37] Uzzau, S., Figueroa-Bossi, N., Rubino, S., Bossi, L. (2001) Epitope tagging of chromosomal genes in *Salmonella. PNAS.* 98,15264-15269.

[38] Malheiros, P.S., Brandelli, A., Tondo, E.C. (2008) Acid and thermal resistance of a *Salmonella* Enteritidis strain involved in several foodborne outbreaks. *Journal of Food Safety*, 29, 302-317.

[39] Perez, K.J., Martins, F.S., Cara, D.C., Nicoli, J.R., Tondo, E.C. (2012) Evaluation of intestinal invasion in germ-free mice challenged with acid-adapted and non acid-adapted *Salmonella* Enteritidis SE86 and *Salmonella* Typhimurium ST99. *Journal of Food Safety*, 32, 108-114.

[40] Scapin, D., Grando, W.F., Rossi, E. M., Perez, K.J. Malheiros, P. da Silva, Tondo, E.C. (2013) Antagonistic activity of *Lactobacillus acidophilus* LA10 against *Salmonella enterica* serovar Enteritidis SE86 in mice. *Brazilian Journal of Microbiology*, 44 (1), 57-61.

[41] Lai, J., Wu, C., Wu, C., Qi, J., Wang, Y., Wang, H., Liu, Y., Shen, J. Serotype distribution and antibiotic resistance of *Salmonella* in food-producing animals in Shandong province of China, 2009 and 2012. (2014). *International Journal of Food Microbiology*, 180, 30–38.

[42] Yang, S.J., Park, K.Y., Kim, S.H., No, K.M., Besser, T.E., Yoo, H.S., Kim, S.H., Lee, B.K., Park, Y.H. (2002) Antimicrobial resistance in *Salmonella* enterica serovars Enteritidis and Typhimurium isolated from animals in Korea: comparison of phenotypic and genotypic resistance characterization. *Veterinary Microbiology*, 86, 295-301.

[43] Breuil, J., Brisabois, A., Casin, I., Armand-lefèvre, L., Frémy, S., Collatz, E. (2000) Antibiotic resistance in salmonellae isolated from humans and animals in France: comparative data from 1994 and 1997. *Journal of Antimicrobial Chemotherapy,* 46, 965-971.

[44] Oliveira, F.A., Pasqualotto, A.P., Padilha, W., Tondo, E.C. (2012) Characterization of *Salmonella* Enteritidis isolated from human samples. *Food Research International*, 1, 2-5.

[45] Paula, C.M.D., Ritter, A.C., Pieta, L., Amaral, P.H., Tondo, E.C. (2011) Antimicrobial resistance in *Salmonella* Enteritidis from foods involved in human salmonellosis outbreaks in Southern Brazil from 2003 to 2006. *Journal of Microbiology and Antimicrobials*, 3, 233-240.

[46] Geimba, M.P., Tondo, E.C., Brandelli, A. (2005) Antimicrobial Resistance in *Salmonella* Enteritidis Isolated from Foods Involved in Human Foodborne Outbreaks Occurred in the South of Brazil, 1999 to 2000. *Journal of Food Safety*, 25, 3, 173-182.

[47] Oliveira, F.A., Brandelli, A., Tondo, E.C. (2006). Antimicrobial resistance in *Salmonell*a Enteritidis from foods involved in human salmonellosis occurred in southern Brazil. *The New Microbiologica*, 29, 49-54.

[48] Malorny, B., Schroeter, A. and Helmuth, R. (1999) Incidence of quinolone resistance over the period 1986 to 1998 in veterinary Salmonella isolates from Germany. *Antimicrobial Agents and Chemotherapy*, 43, 2278–2282.

[49] Threlfall E.J., Day, M., de Pinna, E. (2006) Drug-resistant enteric fever in the UK. *Lancet*, 367:1576.

[50] Marimon, J.M., Goariz, M., Zigorraga, C., Cilla, G. and Perez-Trallero, E. (2004) *Increasing prevalence of quinolone resistance in human nontyphoid* Salmonella enterica *isolates obtained in Spain from 1981 to 2003. Antimicrobial Agents Chemotherapy*, 3, *576–577.*

[51] NCCLS (2005). Clinical and Laboratory Standards Institute/NCCLS. Performance Standards for Antimicrobial Susceptibility Testing, Fifteenth Informational Supplement. CLSI/NCCLS document M100- S15. Clinical and Laboratory Standards Institute, 940 West Valley Road, Suite 1400, Wayne, *Pennsylvania*, 19087-1898 USA.

In: Salmonella
Editor: Christopher B. Hackett
ISBN: 978-1-63463-651-3

Chapter 5

PREVALENCE OF *SALMONELLA* SPP. AND OCCURRENCE OF SALMONELLOSIS FROM READY TO EAT FRESH FRUITS AND VEGETABLES

Agni Hadjilouka, Spiros Paramithiotis*[*] *and Eleftherios H. Drosinos

Laboratory of Food Quality Control and Hygiene,
Department of Food Science and Human Nutrition,
Agricultural University of Athens, Athens, Greece

ABSTRACT

Human food- borne salmonellosis, an illness caused by *Salmonella* species, is an infection of serious public health concern as it is increasing worldwide and appears to be the primary cause of confirmed foodborne outbreaks. It is estimated that tens of millions of cases occur worldwide every year and the disease results in more than hundred thousand deaths. *Salmonella* in food is mainly detected in meat and its products but the numbers of outbreaks caused by vegetables and products thereof have increased compared to previous years. Reported outbreaks caused by consumption of fresh products such as tomatoes, cucumbers, watermelons, cantaloupes, sprouts and juices have been linked to *Salmonella.* Meanwhile *Salmonella* has been isolated from a variety of fresh products and its prevalence has been reported to range from 0.6% to 72% in strawberries, fennel, lettuce, oranges, cantaloupes, onions, sprouts etc. In this chapter prevalence of *Salmonella* in unprocessed and minimally processed fruits and vegetables as well as salmonellosis outbreaks linked to the consumption of such products are addressed.

Salmonella spp. is a microorganism widely spread in poultry and swine and is often distributed in soil surface, decaying vegetation, soil sewage, sewage sludge, municipal solid wastes, animal feces, water, fertilizers and plants (FDA, 2001; Lemunieret al., 2005). It can survive for several weeks in a dry environment, for several months in water and has also the

Corresponding author. E-mail: sdp@aua.gr; Tel.: +30 210 5294705 Fax: +30 210 5294683.

ability to survive in the plant habitat, but not as well as the plant-associated species. Produce contamination and disease outbreaks occur due to the fact that once *Salmonella* is attached to the produce, may maintain in the product during its whole shelf life (Fatica and Schneider, 2011).

Therefore, improper agricultural and manufacturing practices in cultivation, harvest, storage, transport and preparation of the products may result in their contamination and produce-associated outbreaks (Tournas, 2005). Fresh produce are usually consumed either raw, without any undergoing processes, or after minimal process, that causes little microbial reduction.

The nature of the consumption in combination with the fact that most fruits and vegetables are grown in a non-sterile environment and producers can't have full control of the conditions in the field, explain the reason why fresh and minimally processed products may be easily contaminated with pathogens.

Surveys on fresh products have revealed the presence of *Salmonella* spp. in various vegetables and fruits, such as cantaloupes, strawberries, oranges, tangerines, lettuces, cabbages, celeries, parsley and artichokes.

Prevalence of *Salmonella* spp. has been reported to vary from 0.6% (1/180) in scallions to 72% (64/89) in fennels (FDA, 2001; Ercolani, 1976). Generally it appears to be very variable and it can be as low as 0.7% (1/143), 0.76% (11/1440) and 0.9% (1/116) in strawberries, cantaloupes and lettuces, respectively, or even as high as 50% (6/12) and 68% (1/120) in cilantros and lettuces, respectively (FDA, 2001; Madden, 1992; Ercolani, 1976).

PRESENCE OF *SALMONELLA* SPP. IN CATEGORIES OF FRUITS AND VEGETABLES

Prevalence of Salmonella spp. in Raw Fruits

Cantaloupes

According to Madden (1992), in 1990, FDA examined the presence of *Salmonella* spp. in cantaloupes during two surveys in Mexico.

The surveys were performed between March and April, 1990 and between November and January, 1990, respectively. In the first survey 8 *Salmonella* serovars were found in 11 (0.76%) of the 1440 samples and in the second survey 12 (1.1%) of the 2220 samples were positive for the presence of 12 *Salmonella* serovars. Interestingly, melons studied in the second survey originated from the harvest area that was associated with the outbreak of *S.* Chesterin1989-1990 (Table 1).

In March of 1999, FDA started a survey about the presence of the pathogens *Salmonella* and *E. coli* O157:H7 in high volume imported fresh produce. One hundred fifty-one samples of cantaloupes originating from Canada, Costa Rika, Dominican Republic, Ecuador, Guatemala, Haiti, Honduras, Mexico and Nicaragua, and being imported to the U.S. were tested for the presence of *Salmonella* spp. by the FDA. Eight of the samples (5.3%) were positive for the presence of *Salmonella* spp. (FDA, 2001). Similarly, none of the 336 orange and tangerine samples were positive for the presence of the pathogen. Sampling took place in citrus packing houses in the U.S. (Pao et al., 1998).

Table 1. Examples of major reported outbreaks caused by *Salmonella* spp., linked to melons (FDA, 2013)

Pathogen	Year	Location	Source	Venue	Type	No. of cases/deaths	Comments	Reference
Salmonella spp.	1950	Minnesota	-	Roadside stand	Watermelon	6/0	Prepared cut melon. *S.* Bareilly isolated from melon. Melon kept at ambient temperature.	Blostein, (1993)
S. Miami	1954	Massachusetts	Florida	Supermarket	Watermelon	17/1	Organism was isolated from home samples but not from supermarket samples. Melons were from Florida where *S.* Maimi is common.	Gayler et al., (1955)
S. Oranienburg	1979	Illinois	Illinois	Supermarket	Watermelon	18/0	Damaged fruits were cut, covered with plastic film, and displayed, sometimes without refrigeration until sold.	CDC, (1979)
S. Chester	1989-90	Multistate, US	Mexico and Central America	Unknown	Cantaloupe	>245/2	Cut cantaloupe from salad bars.	CDC, (1991); Lund and Snowdon, (2000)
S. Poona	1991	Multistate, US Canada	Texas or Mexico	Unknown	Cantaloupe	>400 confirmed U.S, 72 Canada/0	Fruit salads containing sliced cantaloupes.	CDC, (1991)
S. Javiana	1991	Michigan	-	picnic and school party	Watermelon	26 primary,13 secondary/0	Melon not washed prior to cutting. Suspected contamination from melon rind. Melon served over 3 hour period at room temperature. Leftovers served the next day.	Blostein, (1993)
S. Saphra	1997	California	Mexico	Home, grocery stores, restaurants	Cantaloupe	24/0	Probable contamination during production or harvest. Lack of refrigeration at retail might contributed to the outbreak.	Mohle-Boetani et al., (1999); FDA, (2013)
S. Oranienburg	1998	Ontario, Canada	U.S, Mexico or Central America	Various	Cantaloupe	22/0	Possible contamination with organism from surface when slicing. Cut fruit was probably left sitting at room temperature for several hours before consumption.	Deeks et al., (1998)
S. Poona	2000	Multistate, US (8 states)	Mexico	Various	Cantaloupe	NR/NR	-	FDA, (2013)

NR= Not Reported.

Strawberries

During FDA's survey in 1999, a total of 143 strawberry samples coming from Argentina, Belgium, Canada, Mexico and New Zealand was collected. Only one (0.7%) of the samples was positive for the presence of the pathogen (FDA, 2001).

Oranges - Tangerines

None of the 375 orange samples were positive for the presence of *Salmonella* in the survey performed in orchards through juice plants in the U.S. Fruit surfaces and juice were analyzed and 3 categoriesof orange samples were tested: graded hulls, washed and graded and ungraded (FDA, 2013).

Prevalence of Salmonella spp. in Raw Vegetables

Seed Sprouts

Three hundred forty-four mung bean samples were collected from open markets in Thailand within 7 months and examined for *Salmonella* presence. Thirty of the collected samples were positive and prevalence was 8.7% (Jerngklinchan and Saitanu, 1993). On the contrary, none of the 13 samples collected in survey performed within 6 months in the U.S was positive for *Salmonella* (Splittstoesser et al., 1983).

Lettuces

Lettuces imported into the U.S. from Belgium, Canada, Chile, France, Guatemala, Israel, Italy, Korea, Mexico, Netherlands and Peru were tested by the FDA in 1999. The final prevalence of *Salmonella* spp. was 0.9% (1/116) (FDA, 2001).

In 1976, Ercolani reported 68% *Salmonella* spp. prevalence in lettuce samples. One hundred twenty samples were purchased from five retail outlets in Italy and they were collected at monthly intervals for one year (Ercolani, 1976).

Lettuce samples from various places in the Netherlands were also tested for the presence of *Salmonella* spp. Samples were firstly tested for the presence of *E. coli* and further investigation for the presence of *Salmonella* was performed only in the samples that were *E. coli* positive. Two of the 28 samples were finally positive for *Salmonella* and prevalence was 7.1% (Tamminga, 1978).

Eighty samples of lettuce were tested in Spain. Samples were collected from farms and wholesale and retail markets. Presence of the pathogen was attributed to possible use of contaminated irrigation water. Two of the 80 samples were positive and prevalence was estimated at 6.3% (Garcia-Villanova Ruiz et al., 1987a).

On the other hand, none of the 151 imported lettuce samples collected in retail markets of the U.K. was positive for *Salmonella* spp. (Little et al., 1999).

Beet Leaves

Twenty-six beet leave samples were collected from various places in Spain within one year. Four (7.7%) of the samples were found positive for *Salmonella* spp. and contamination was attributed to contaminated irrigation water (Garcia-Villanova Ruiz et al., 1987b).

Artichokes and Cardoons

Garcia-Villanova Ruiz et al. (1987b) examined 25 artichoke and 4 cardoon samples collected from a farm and wholesale and retail markets in Spain. Samples were collected within one year and analysis showed that 3 (12%) artichokes and 1 (20%) cardoon samples were positive for *Salmonella* spp. It was considered that it was due to contaminated irrigation water.

Endive

Endive samples either grown in the Netherlands or imported from Italy into the Netherlands were tested for *Salmonella* presence. Twenty-six samples were collected and 2 (7.7%) of them were positive (Tamminga et al., 1978).

Cabbage

Cabbage samples were collected by various places in Spain and tested for *Salmonella* presence. A total of 41 samples was purchased and 7 (17%) of them were positive for pathogen's presence (Garcia-Villanova Ruiz et al., 1987b).

On the other hand, according to Tamminga et al. (1978) none of the 18 cabbage samples collected from various places in the Netherlands was positive for *Salmonella* presence.

Eggplant

Thirteen eggplant samples were tested for the presence of *E. coli* and further investigation for the presence of *Salmonella* spp. was performed only in the samples that were *E. coli* positive. Two (1.5%) samples were finally positive for *Salmonella* spp. (Tamminga et al., 1978).

Cauliflower

Thirteen cauliflower samples from the Netherlands were examined for the presence of *E. coli* and *Salmonella* spp. Presence of *Salmonella* was tested only in positive for the presence of *E. coli* samples and prevalence was finally estimated at 7.7% (Tamminga et al., 1978).

Garcia-Villanova Ruiz et al (1987b) analyzed 23 cauliflower samples from farm and wholesale and retail markets in Spain. One (4.3%) of them was contaminated with *Salmonella* spp. and it was linked to contaminated irrigation water.

Spinach

Thirty eight spinach samples were examined for *Salmonella* presence in Spain. Samples were collected from various places and the contaminated ones were thought to be infected due to contaminated irrigation water (Garcia-Villanova Ruiz et al., 1987b).

Celery

None of the twenty celery samples collected from various places in the Netherlands was found positive for *Salmonella* spp. (Tamminga et al., 1978).

Contrary to the above result, 1.2% prevalence was estimated in 84 celery samples that were imported in the U.S. and were collected from Canada and Mexico (FDA, 2001). In addition, 2/26 (7.7%) celery samples, collected from a farm and wholesale and retail markets in Spain, were positive for *Salmonella* presence (Garcia-Villanova Ruiz et al., 1987b).

Fennel

Ercolani (1976) estimated the occurrence of *Salmonella* in fennel samples from 5 retail outlets in Italy. Sampling was performed on a monthly basis for one year. A total of 89 samples was collected and 64 (72%) of them were contaminated with *Salmonella* spp.

Examination of 15 fennel samples was also conducted in the Netherlands. Samples were either grown in the Netherlands or imported from Italy. None of samples was found positive for *Salmonella* presence (Tamminga et al., 1978).

Scallions

One hundred eighty scallion samples from Canada, Guatemala, Israel, Italy and Mexico were tested for *Salmonella* presence. Scallions were imported into the U.S. from the above countries and sampling took place in various places. One sample was contaminated and prevalence was 0.6% (FDA, 2001).

Zucchini & Peppers

None of the 20 pepper and 11 zucchini samples examined in the Netherlands was positive for the presence of *Salmonella* spp. (Tamminga et al., 1978).

Prevalence of Salmonella spp. in Mixed Raw Vegetables

Rude et al. (1984) conducted a 2-year survey in order to estimate the contamination by nematodes, amoebae, and bacteria of the genus *Salmonella* in salad vegetables collected from wholesale and retail sources in the U.S. The vegetables examined were cucumbers, cabbage, lettuce, celery, carrots, radishes, tomatoes, mushrooms, cauliflower, and spinach. Four samples were found positive for the presence of *Salmonella* spp.

Seven percent prevalence of *Salmonella* spp. was reported in mixed raw vegetables in Iraq. Forty-three samples were collected from various places in Baghdad and 3 of them were positive for the pathogen's presence (Al-Hindawi et al., 1979).

Mixed raw vegetables and irrigation water samples were collected by various places in Spain. Eight hundred forty-nine samples were tested and 46 (5.4%) of them were positive for *Salmonella* spp. Results indicated that *Salmonella* spp. isolates from produce had a close relationship to the ones that had been isolated by water samples (Garcia-Villanova Ruiz et al., 1987b).

PREVALENCE OF *SALMONELLA* SPP. IN GREEN AND MIXED SALADS

One hundred fifty-nine mixed salad and 57 green salad samples were purchased from retail markets and food services in Egypt. *Salmonella* spp. was isolated from 2 green salad samples and 1 mixed salad sample and prevalence was 3.5% and 0.6%, respectively (Saddik et al., 1985).

Lin et al. (1996) examined the occurrence of *Salmonella* in 63 mixed salad samples in the U.S. Samples were purchased from 31 food service facilities (4 supermarkets, 14 fast food chain restaurants, and 13 family restaurants and none of them was positive for *Salmonella* spp.

PREVALENCE OF *SALMONELLA* SPP. IN RAW HERBS AND SPICES

Parsley

Twenty three samples of parsley were purchased from a farm, wholesale and retail markets in Spain. Samples were collected within one year and examination showed prevalence 4.3%. It was suggested that contamination occurred due to contaminated irrigation water (Garcia-Villanova Ruiz et al., 1987a).

Eighty-four parsley samples that were imported into the U.S. were collected from Canada, Columbia, Costa Rica, Israel, Italy, Mexico and Peru and examined for *Salmonella* presence. One of them was positive and prevalence was 1.2% (FDA, 2001).

Chili

Prevalence of 31% was estimated in two different surveys performed in chili samples. Sixteen chili samples were collected in the Netherlands and in Surinam and in both cases 4 of the samples were positive for the presence of *Salmonella* spp. (Tamminga et al., 1978; Tauxe et al., 1997).

Cilantro

One hundred seventy-seven samples of cilantro that were imported into the U.S. were collected from Canada, Costa Rica, Mexico, Peru and Trinidad & Tobago. Sixteen of the samples were positive for *Salmonella* spp. and prevalence was 9% (FDA, 2001).

Culantro

Twelve samples of culantro that were imported into the U.S. were collected from Costa Rica and Trinidad & Tobago. Six of the samples were positive for *Salmonella* spp. and prevalence was 50% (FDA, 2001).

SALMONELLOSIS OUTBREAKS ASSOCIATED WITH FRUITS AND VEGETABLES

Salmonella spp. has been reported as a source of illnesses since the late 1800s (Fatica and Schneider, 2011). The onset of the illness symptoms occurs 6-72 hours after the intake of the pathogen and more usually occurs between 12 and 36 hours. Infection's symptoms are usually acute fever, abdominal pain, diarrhea, nausea and sometimes vomiting. It usually lasts 2-7 days and due to the fact that symptoms of salmonellosis are relatively tender, the infection is most of the times self-limiting and patients recover without any treatment. However, severe infections can occur and especially when the patients are elderly, immune-compromised or young children. In these cases the dehydration caused by the illness can become very serious and even life- threatening (WHO, 2013).

Table 2. Examples of major reported outbreaks caused by *Salmonella* spp., linked to seeds (FDA, 2013)

Pathogen	Year	Location	Seed Source	Type of sprout	No. of cases/deaths	Comments	References
S. Virchow	1988	U.K	Thailand and Australia	Mung	7/0	Probably from the same outbreak as *S.* Saintpaul in UK.	FDA, (2013)
S. SaintPaul *S.* Havana, *S.* Muenchen	1988	Sweden	NR*	Mung	148/0	Probably same seeds as UK outbreak. *S.* Havana and *S.* Muenchen but not *S.* Saintpaul	Nguyen-the and Carlin (2000); FDA, (2013)
S. Saintpaul	1988	U.K.	Thailand and Australia	Mung	143/0	Multiple serovars isolated from bean sprouts, seeds, and environmental samples (from producer waste materials).	FDA, (2013)
S. Gold-Coast	1989	U.K	The Netherlands	Cress	31/0	Contaminated seed and/or sprouter.	Feng, (1997); Taormina et al., (1999)
S. Bovismorbificans	1994	Sweden and Finland	Australia	Alfalfa	595/0	Contaminated seeds from the same seed lot and importer.	Ponka et al., (1995)
S. Newport	1995	Denmark (probably U.S and Canada)	The Netherlands	Alfalfa	154/0	Source of contamination unknown.	Feng, (1997); FDA, (2013)
S. Stanley	1995	Multistate U.S, Canada and Finland	The Netherlands	Alfalfa	>272/0	Seeds from the same sprouter. At least 4 seed lots involved. Possible contamination occurred prior to shipping.	Mahon et al., (1997)
S. Newport	1995	British Columbia, Canada, Oregon (probably Georgia and Vermont) and Denmark	The Netherlands	Alfalfa	133/0	Seeds from the same sprouter. At least 4 seed lots involved. Possible contamination occurred prior to shipping.	Feng, (1997); Taormina et al., (1999)
S. Montevideo; *S.* Meleagridis	1995-96	California	California	Alfalfa	>500/1	The sprouts were traced to a specific sprouter. Seeds traced to single California seed grower.	Taormina et al., (1999); FDA, (2013)
S. Meleagridis	1996	Canada	Unknown	Alfalfa	124/0	Sprouts were organically grown with no chlorine pre-soak.	Feng, (1997); FDA, (2013)
S. Infantis; *S.* Anatum	1997	Kansas and Missouri	Unknown	Alfalfa	109/0	Seeds were believed to be contaminated.	Feng, (1997); Taormina et al., (1999)
S. Senftenberg	1997	California	5 states	Alfalfa and clover sprouts	52/0	Contaminated seeds from to a specific sprouter suspected. Same sprouter as 1998 *E. coli* O157: NM outbreak.	FDA, (2013); Taormina et al., (1999)

Pathogen	Year	Location	Seed Source	Type of sprout	No. of cases/deaths	Comments	References
S. Havana; *S.* Cubana; *S.* Tennessee	1997-98	California	California	Alfalfa	34/0	Suspected contaminated seeds	FDA, (2013); Taormina et al., (1999)
S. Havana	1998	California and Arizona	NR	Alfalfa	14(California) 4 (Arizona)/1	Sprouts traced to a single producer. Seeds obtained from the same lot yielded sprouts from which *S.* Havana was cultured.	Backer et al., (2000)
S. Mbandaka	1999	Oregon, California, Idaho, and Washington	California	Alfalfa	Approx. 68/0	Seeds were believed to come from the same lot and distributed to various growers in California, Florida, and Washington. No cases in Florida.	FDA, (2013); Taormina et al., (1999)
S. Paratyphi B var. Java	1999	Alberta, British Columbia, and Saskatchewan, Canada	Unknown	Alfalfa	46/0	Spouts originated from the same brand or common seed source.	FDA, (2013)
S. Enteriditis	2000	Alberta and Saskatchewan, Canada	China	Alfalfa	8/0	Outbreaks occurred at 5 Vietnamese restaurants. Sprouts came from 2 growers who received seeds imported from China.	FDA, (2013)
S. Enteriditis	2000	California	China	Mung	45/0	Cluster of illness linked to 3 Vietnamese restaurants. *S.* Enteriditis isolated from environment at sprouter.	FDA, (2013)

NR= Not Reported.

Table 3. Examples of major reported outbreaks caused by *Salmonella* spp. linked to tomatoes and mamey (FDA, 2013)

Pathogen	Year	Location	Produce Source	Venue	Type of Produce	No. of cases/deaths	Comments	Reference
S. Javiana	1990	Multistate, U.S.	South Carolina	Various	Tomatoes	174/0	Contamination of water bath used by packer.	Tauxe, (1997); Beuchat, (1996)
S. Montevideo	1993	Multistate, U.S.	South Carolina	Various	Tomatoes	84/0	Contamination of water bath used by packer.	Lund and Snowdon, (2000); Wei et al., (1995); Tauxe, (1997)
S. Typhi	1998-99	U.S.	Brazil	Unknown	Mamey	13/0	Source of contamination not known.	Lund and Snowdon, (2000)
S. Baildon	1998-99	Multistate, U.S.	Florida	Various	Tomatos	85/3	Possible contamination by domesting or wild animals.	Susman, (1999); FDA, (2013)

Table 4. Examples of major reported outbreaks caused by *Salmonella* spp. linked to unpasteurized fruit juice (FDA, 2013)

Pathogen	Year	Location	Type of Juice	Venue	No. of cases/deaths	Comments	Reference
S. Typhi	1989	France	Apple	NR	NR/NR	-	FDA, (2013)
S. Typhi	1922	France	Apple	NR	23/0	Apples washed with non-potable water	Paquet, (1923)
S. Typhimurium	1974	New Jersey	Apple	Farm and small retail outlets	296/0	Juice made by a high proportion of dropped apples. Fertilization of apple trees with manure. Equipment rinsed with cold, not sanitized water. Six of thirty employees	CDC, (1975)
S. Gaminera; *S.* Hartford; *S.* Rubislaw	1995	Florida	Orange	Retail	62 ill and 7 hospitalized/0	In-plant sanitation problems found. Surface water was used for orchard irrigation. Drops were used for juice. *Salmonella* was isolated from amphibians and soil around the processing plant.	CDC, (1995); Cook et al., (1998)
S. Muenchen	1999	Mexico	Orange	Restaurant	207 confirmed, 91 suspected/1	Multiple strains of *Salmonella* collected from producer. Juice squeezed in Mexico and transported to Arizona in tanker trucks where it was bottled. Ice was added illegally to juice prior to transport.	CDC, (1999)
S. Typhimurium	1999	Australia	Orange	Retail	405/0	*Salmonella* was isolated from unopened cartons of orange juice.	FDA, (2013)
S. Enteriditis	2000	California	Citrus	Retail and food services	14/0	Gallon sized containers of citrus juices were implicated in the outbreak.	Buttler, (2000)
S. Typhimurium, *S.* Saintpaul	2005	-	Orange	Company	152/0	Company non-compliant with HAACP regulation	Jain et al., (2009)

NR= Not Reported.

Table 5. Examples of major reported outbreaks caused by *Salmonella* spp. linked to reconstituted orange juice (FDA, 2013)

Pathogen	Year	Location	Produce	Type of juice	Venue	No. of Cases/deaths	Comments	Reference
S. Typhi	1944	Ohio	Unknown	Orange	Residential 1 hotel	18/1	Juice was handled by an asymptomatic food worker.	Duncan et al., (1946)
S. Typhi	1989	New York	Unknown	Orange	Resort Hotel	46 confirmed, 24 suspected/0	Utensils used were difficult to clean. Orange juice distributed near restrooms.	Birkhead et al., (1993)

In addition, the infection may spread from the intestines to the blood stream and then to other body sites and can lead to death, unless the patients are treated with the appropriate antibiotics; in these cases, patients need to be hospitalized. *Salmonella* infection can also cause, in rare occasions, a long-term disease called reactive arthritis. Patients develop pain in their joints, irritation of the eyes and painful urination. The symptoms can last for months or even years and can lead to chronic arthritis which is difficult to treat (CDC, 2010d). It is generally believed that 60 to 80% of the salmonellosis cases are not recognized as part of an outbreak and they are considered as sporadic cases, or even that they are not reported or diagnosed at all (WHO, 2013).

The genus *Salmonella* is divided into two species, *S. bongori* and *S. enterica*. *S. enterica* is divided into more than 2,500 distinct serovars (Fatica and Schneider, 2011). *S. enterica* serovar Enteritidis and *S. enterica* serovar Typhimurium are the most common serovars and are associated with over the half of the reported salmonellosis cases both in the European Union and the United States (CDC, 2010; EFSA and ECDC, 2012; Fatica and Schneider, 2011).

In the European Union, *Salmonella* is the main cause of confirmed foodborne outbreaks and the second most reported zoonosis, after *Campylobacter* (EFSA and ECDC, 2012). In the United States nontyphoidal *Salmonella* spp. is the second cause of illness after Norovirus and the first cause of hospitalizations and deaths (Scallan et al., 2011). It is estimated that the pathogen causes about 6.2 million and 1 million illnesses in the EU and the United States, respectively, each year (Havelaar et al., 2012; Scallan et al., 2011).

Before 1990, most of the infections caused by *Salmonella* spp. were connected to the consumption of contaminated poultry and products thereof (Tauxe et al., 1997). However, the number of cases linked to vegetables and vegetable products have significantly increased the past years and *Salmonella* spp. has been linked to a large number of outbreaks caused by the consumption of these products, such as melons (Table 1), seed sprouts (Table 2), raw tomatoes (Table 3) and unpasteurized and reconstituted juices (Table 4 & 5). CDC estimates that 95% of *Salmonella* infections are caused by food-borne produce-based sources (Fatica and Schneider, 2011). One of the most important reasons that influence the increase of salmonellosis due to vegetables, are the changes in agricultural practices that globalize the food supply and disseminate the pathogens over wide geographical areas. Moreover, changes in eating habits that increased the popularity of the fresh products - due to consumers' tendency to eat healthier- have increased the commerce of fresh produce (Collins, 1997).

Major Outbreaks of Salmonella spp. in the United States and Canada

Multistate Outbreak of *S.* Typhimurium Infections Linked to Tomatoes, 2006

A multistate outbreak caused by *S.* Typhimurium was reported the autumn of 2006. Analyses of the data collected by investigators indicated that responsible for the outbreak was the consumption of tomatoes served in restaurants. The majority of the patients became ill the last two weeks of September, 2006. Among the 111 ill persons who provided the date of the beginning of their illness, 93% became ill between September 14 and October 2. The outbreak caused 183 cases totally, in 21 states: Alabama (1), Arkansas (4), Connecticut (28), Georgia (1), Indiana (1), Kentucky (19), Massachusetts (50), Maine (8), Michigan (2), Minnesota (14), North Carolina (4), Nebraska (1), New Hampshire (14), Ohio (4), Pennsylvania (3), Rhode Island (6), Tennessee (9), Virginia (3), Virginia (8), Washington (1), and Wisconsin (2). All the patients resided east of the Mississippi River except for a Washington state resident who had traveled to the Northeast U.S. In addition, 2 patients infected by the outbreak of *S.* Typhimurium had been reported from Canada, one of whom had previously traveled to an affected U.S. state. The median age of patients was 34 years old and 57% were female. Most of them had fever and diarrhea and no deaths were reported. Finally, according to people from whom information was available, 22 patients (12%) were hospitalized (CDC, 2006).

Multistate Outbreak of *S.* Wandsworth Infections Linked to Veggie Booty, 2007

In 2007 public health authorities investigated a multistate outbreak caused by *S.* Wandsworth, a rare *Salmonella* serovar. Since 18th of July 2007, 65 infected people have been reported to CDC from 20 states: California, Colorado, Connecticut, Georgia, Illinois, Indiana, Massachusetts, Maryland, Minnesota, New Hampshire, New Jersey, New York, Oregon, Pennsylvania, Tennessee, Texas, Vermont, Virginia, Washington, and Wisconsin. Among the patients for whom clinical information was available six were hospitalized, 76% had bloody stools and all of them had diarrhea. No deaths were reported. Sixty-four cases (98%) occurred between 26th of February and 27th of June, 2007 and most of them (91%) occurred in children aged from 10 months to 3 years old. A multi-state case-control study demonstrated that the consumption of Robert's American Gourmet brand "Veggie Booty" -a snack of puffed rice and corn with a vegetable coating- was the source of the illness and led to the multistate outbreak. On 28th of June 2007, the company that manufactured the product, after being informed about the outbreak by FDA, issued a voluntary recall. After the recall of these products no illness from *S.* Wandsworth was reported (CDC, 2007).

The FDA laboratory and the New York State Department of Health Wadsworth Center Laboratory isolated the outbreak strain from sealed bags of Veggie Booty. Furthermore, the Minnesota Department of Agriculture Laboratory (MDAL) isolated *S.* Wandsworth and *S.* Typhimurium from sealed bags of Veggie Booty obtained from retail stores. Additionally, *S.* Typhimurium caused infection in 10 people between 1st and 27th of June, 2007. Eight of the patients were interviewed by OutbreakNet officials and it was determined that all of them had consumed Veggie Booty during the week before their illnesses began.

Multistate Outbreak of *S.* Litchfield Infections Linked to Cantaloupe, 2008

In 2008 a multistate outbreak caused by consumption of cantaloupes was connected to *S.* Litchfield. Responsible for this outbreak were determined to be some cantaloupes from a Honduran company called "Agropecuaria Montelibano". As of the 2nd of April 2008, 51 ill people were identified in 16 states: Arizona (1), California (10), Colorado (1), Georgia (2),

Illinois (1), Missouri (1), New Jersey (2), New Mexico (1), New York (5), Ohio (1), Oklahoma (2), Oregon (5), Tennessee (1), Utah (5), Washington (10), and Wisconsin (3). Onset dates, which are known for 50 patients, ranged from 10th January to 10th March, 2008. Their ages ranged from <1 to 93 years old and 59% of them were female. At least 16 persons had been hospitalized but no deaths were reported. In addition, 9 ill persons with the outbreak strain were reported in Canada (CDC, 2008a).

Multistate Outbreak of *S.* Saintpaul Infections Linked to Raw Produce, 2008

On 22nd May 2008, the New Mexico Department of Health notified CDC that 4 people were infected by *S.* Saintpaul and that15 other people had *Salmonella* infection whose isolates had not yet been characterized.

The cases of illness caused by the outbreak strain continued to be reported and the outbreak expanded in 43 states, the District of Columbia and Canada. Five ill people were reported in Canada, 4 appeared to have been infected while traveling in the United States while the travel status of the fifth ill person was unknown. The number of ill people identified in each state was: Alabama (8), Arkansas (21), Arizona (59), California (16), Colorado (17), Connecticut (5), Florida (4), Georgia (42), Idaho (6), Illinois (120), Indiana (21), Iowa (2), Kansas (21), Kentucky (2), Louisiana (3), Maine (1), Maryland (39), Massachusetts (31), Michigan (28), Minnesota (31), Mississippi (2), Missouri (20), Montana (1), New Hampshire (6), Nevada (14), New Jersey (16), New Mexico (115), New York (41), North Carolina (28), Ohio (10), Oklahoma (38), Oregon (11), Pennsylvania (15), Rhode Island (3), South Carolina (2), Tennessee (10), Texas (559), Utah (3), Virginia (31), Vermont (2), Washington (18), West Virginia (1), Wisconsin (13), and the District of Columbia (1). The investigation held by the public health authorities in several states with the assistance of CDC, showed that jalapeno peppers were a major source of contamination and that serrano peppers were also source of the outbreak. In addition, tomatoes were possibly a source, particularly early in the outbreak. Jalapeno peppers were traced to be grown and packed in Mexico and then be distributed in the United States. The outbreak strains were isolated from jalapeno peppers in a U.S. warehouse and a patient's home and from serrano peppers and water that was collected from a farm in Mexico. *S.* Saintpaul had caused the infection of 1442 people by the end of the outbreak.

Based on the 1414 patients with clinical information available, illnesses began between 16th April and 11th August, 2008 with the most people becoming ill during May or June; 119 became ill on 1st July or later.

The latest onset day was reported to be 11th August 2008. Patients ranged in age from < 1 to 99 years old with a median age of 33 years old. The rate of illness was highest among young people from 20 to 29 years old.

At least 286 people were hospitalized, and the infection might have contributed to two deaths. The previous rarity of this strain and the distribution of illnesses in all U.S. regions suggested that the implicated food had been distributed throughout much of the country. In addition, the fact that a lot of people with *Salmonella* illness do not have a stool specimen tested makes it likely that many more illnesses had occurred than those that were reported (CDC, 2008b).

Multistate Outbreak of *Salmonella* Infections Linked to Pistachio Nuts, 2009

On 26th of March, 2009 the US Food Administration (FDA) informed CDC that multiple samples of pistachio nuts and pistachio-containing products were contaminated with several serotypes of *Salmonella*, including Montevideo, Newport, and Senftenberg. The products had been collected and tested over several months from a single company called "Setton Pistachio of Terra Bella Inc, California". The company stopped the distribution of the raw shelled, roasted shelled and roasted-in-shell pistachios from its 2008 crop and issued a voluntary recall of these products. In addition, the companies that were supplied with pistachios by Setton Pistachio recalled also their products. It is also likely that contaminated pistachios had been used in a wide range of foods, including cakes, cookies, puddings, trail mix, snack bars, and ice cream (CDC, 2009a).

Multistate Outbreak of Human *S.* Saintpaul Infections Linked to Raw Alfalfa Sprouts, 2009

In February 2009, the Nebraska Department of Health Human Services identified a higher than expected number of *Salmonella* serotype Saintpaul isolates from sick people. This led to a series of epidemiologic investigations by state and local health officials. The cases were divided into two groups, the confirmed and the probable ones. As confirmed cases were defined the patients whose stool culture on or after February 1st, 2009 was positive for the presence of *S.* Saintpaul with the outbreak strain PFGE patterns. On the other hand, probable cases were characterized from the patients whose stool culture was positive for the presence of *S.* Saintpaul with a pending PFGE pattern. On 26th February, 2009 a notification about a group of case-patients infected by *S.* Saintpaul among residents of Nebraska was distributed to US State public health officials. Additional cases were then reported from health departments in Iowa, South Dakota, Minnesota, Kansas and Missouri. Five of the 14 Nebraska case-patients stated the same restaurant chain (Chain A) and 9 of them had recently consumed alfalfa sprouts, while from the first 7 Iowa case- patients 1 had visited Chain A and 6 had consumed alfalfa sprouts. Alfalfa sprouts was the most common food that case patients had previously consumed and no other food was significantly associated with the illness. In addition, study of the patients from Nevada and Iowa showed that the case-patients were more likely to have consumed alfalfa sprouts than controls (27/32 vs 5/32) and that case patients were significantly more likely to have eaten at Chain A than controls (24/32 vs 10/32). By 19th March, 2009, 186 cases had been identified and of the 156 patients with the available medical information, 114 (73%) reported alfalfa sprout consumption. Further interviews and studies indicated that although there was a variety of companies distributing the restaurants and stores reported by the patients, alfalfa sprouts originated from the same sprouting facility (Facility A) in Omaha, Nebraska. One hundred twelve patients out of the 114 (98%) with sprout exposure were linked to a restaurant or a retail outlet that had received the sprouts from Facility A. On 3rd March, 2009, Facility A voluntarily recalled the products. Assessment of the shipment, production and distribution dates suggested that the outbreak was correlated with the distribution of sprouts from a seed shipment that arrived as Facility A on 13th January, 2009 and was last sprouted on 13th, 2009. Due to the fact that during the period of the outbreak multiple lots of seed were purchased only from one seed company

"Seed Company B" for the production of alfalfa sprouts, all the lots of seeds originated from the same seed grower (Grower C) were identified with the prefix "032". FDA's investigation showed that the lots of seeds identified with the prefix "032" were sold by Seed Company B in many states and in many facilities. For this reason and due to the fact that cases of illness were spread across multiple states, on 26th April, 2009, FDA and CDC recommended that consumers should not eat raw alfalfa sprouts, including sprout blends containing alfalfa sprouts, until further notice, while on 1st May, 2009 Seed Company B voluntarily withdrew all alfalfa seeds bearing six digit lot numbers that started with "032".

Until 7th May, 2009, 235 people from 14 states had been reported to be infected with the outbreak strain of *S.* Saintpaul. The number of patients identified in each state is as follows: Nebraska (111), Iowa (35), South Dakota (38), Michigan (19), Kansas (8), Pennsylvania (7), Minnesota (5), Ohio (3), Illinois (2), Virginia (2), West Virginia (2), Florida (1), North Carolina (1), and Utah (1). Among the 234 persons with known illness onset dates, illnesses began between 1st February and 15th April, 2009. Patients aged from < 1 to 85 years old and 68% were female. No deaths were reported and according to the patients with available medical information 3% were hospitalized (CDC, 2009b).

Multistate Outbreak of Human *S.* Montevideo Infections Linked to Black and Red Pepper, 2010

Salmonella has over 2,500 serotypes. Serotype Montevideo is in the top 10 of the most common *Salmonella* serotypes.

The fact that the outbreak strain of *S.* Montevideo is the most common pattern of Montevideo serotype makes the detection of an outbreak challenging. CDC in collaboration with the public health officials in many states, the U.S. Department of Agriculture's Food Safety and Inspection Service (FSIS), the U.S. Food and Drug Administration (FDA), and the State of Rhode Island investigated a multistate outbreak caused by *S.* Montevideo in 2009-2010.

Initial studies that took place between the 16th -21st January, 2010, comparing foods eaten by 41 ill and 41 healthy people suggested that salami was the possible source of the illness. Fifty eight percent of the ill people reported having consumed salami. Sixteen ill people stated that they had purchased the same type of sliced salami variety pack from different grocery store locations before becoming ill.

In addition, 3 ill people who had purchased a similar type of sliced salami deli tray before becoming ill were identified.

All these variety packs and deli trays contained salami made with black pepper, which was added after the lethality step. After the suggestion that these products are responsible for the outbreak, the company Daniele International Inc. recalled the sliced variety pack and the sliced salami deli trays. Audits conducted by the Rhode Island Department of Public Health in the company Daniele International Inc. showed that black and red pepper, intended for the production of Italian-style meats, were positive for the presence of the pathogen *S.* Montevideo. Since then, several recalls had been issued.

During the multistate outbreak, Illinois Department of Public Health carried out product checks that resulted in positive findings, as far the presence of the microorganism was concerned. As a result of the finding of *S.* Montevideo in an unopened salami product the recalls were expanded. Subsequent testing revealed that the product contained also the *S.* Senftenberg strain.

The company stated that black pepper was a possible source of *Salmonella* contamination. On 28th of January, 2010, the Rhode Island Department of Health announced that an opened container of black pepper used in the manufacturing of at least some of the recalled products were positive for *S.* Montevideo and that the DNA fingerprint matched the outbreak strain. On 16th February, 2010, Salame Panino products, that contained crushed red pepper, were found to be positive for the presence of the pathogen. This finding identified that crush red pepper was another source of contamination at Daniele International Inc.

The onset day of the infection was reported to be the 1st July, 2009. As of 28th April, 2010, 272 people from 44 states and the District of Columbia were infected with the outbreak strain. The number of ill people identified in each state was: Alaska (1), Alabama (2), Arizona (9), California (32), Colorado (5), Connecticut (5), District of Columbia (1), Delaware (3), Florida (7), Georgia (3), Iowa (1), Idaho (4), Illinois (28), Indiana (4), Kansas (5), Louisiana (2), Massachusetts (16), Maryland (1), Maine (1), Michigan (4), Minnesota (7), Missouri (3), Mississippi (1), North Carolina (11), North Dakota (1), Nebraska (3), New Hampshire (2), New Jersey (9), New Mexico (2), New York (20), Ohio (9), Oklahoma (1), Oregon (10), Pennsylvania(7), Rhode Island (2), South Carolina (1), South Dakota (4), Tennessee (5), Texas (7), Utah (9), Virginia (1), Washington (19), Wisconsin (1), West Virginia (1), and Wyoming (2). According to the people with reported dates available, illness began on 1st of July, 2009 and the most recent reported onset date was on 14th of April, 2010. Patients' age ranged from <1 year old to 93 years old and the median age was 37 years old.

Fifty-three percent of the patients were male and from the 203 patients with available clinical information, 52 (26%) were hospitalized with no reported death (CDC, 2010a).

Multistate Outbreak of Human *S.* Newport Infections Linked to Raw Alfalfa Sprouts, 2010

An outbreak of *S.* Newport that was connected with the consumption of alfalfa sprouts was reported in 11 states in 2010. From 1st March, 2010, until 24th June, 2010, a total of 44 infected people was reported. Analytically, the number of people reported in each state was: Arizona (4), California (19), Colorado (1), Idaho (6), Illinois (1), Missouri (2), New Mexico (2), Nevada (4), Oregon (2), Pennsylvania (1), and Wisconsin (2). According to patients' available information the symptoms started to occur between 1st March and 1st June, 2010, and 7 of them (19%) were hospitalized. No deaths were reported. Patients ranged in age from <1 to 85 years old with the median age of 38 years old. Sixty-eight percent of patients were female. Interviews of the patients showed that most of them reported eating raw alfalfa sprouts before becoming ill. Some of them had consumed the sprouts at restaurants and others reported purchasing the sprouts from grocery stores.

The implicated sprouts were finally traced in a single sprout processor in California. On 21st May, 2010, J.H. Caldwell and Sons Inc. of Maywood, CA, voluntarily recalled several brands of alfalfa sprouts distributed to wholesale distributors, restaurants, delicatessens, and grocery stores (CDC, 2010b).

Multistate Outbreak of Human *Salmonella* I 4,[5],12:i:- Infections Linked to Alfalfa Sprouts, 2010

Since 1996, there have been at least 30 reported outbreaks of foodborne illness associated with different types of raw and lightly cooked sprouts. Most of these outbreaks were caused by *Salmonella* and *E. coli* infections. From 1st November, 2010, through 9th February, 2011, 140 people from 26 states and the District of Columbia were infected by the outbreak strain of *Salmonella* I 4,[5],12:i:-. The number of ill people identified in each state was: Arkansas (1), California (1), Colorado (1), Connecticut (1), District of Columbia (1), Georgia (1), Hawaii (1), Iowa (1), Illinois (70), Indiana (13), Kentucky (1), Louisiana (1), Massachusetts (2), Maryland (1), Missouri (23), Nebraska (1), Nevada (1), New Jersey (1), New York (2), North Carolina (1), Oregon (1), Pennsylvania (4), South Carolina (1), South Dakota (1), Tennessee (2), Virginia (2), and Wisconsin (4). From the 138 people for whom clinical information was available, the onset dates of the illness ranged from 1st November, 2010, to 18th January, 2011. Twenty-four percent of the patients were hospitalized and no deaths were reported. Patients' age ranged from 1 to 85 years-old, with the median age of 28 years-old. Eighty-seven patients (63%) were female. Nevertheless, some of the cases identified might not be related to this outbreak, due to the fact that the pulsed-field gel electrophoresis (PFGE) pattern associated with this particular *Salmonella* serotype commonly occurs in the United States.

Results of the investigation indicated a link between the infection and the consumption of Tiny Greens Alfalfa Sprouts or Spicy Sprouts (which contained alfalfa sprouts mixed with radish and clover sprouts). The sprouts were distributed to various customers, including farmers' markets, restaurants, and groceries in Illinois, Indiana, Iowa, and Missouri. Sprouts might have also been distributed to other Midwestern states. Half of the illnesses occurred in Illinois, where many of the patients had previously consumed sandwiches containing sprouts at various Jimmy John's outlets. On 29th December, 2010, Tiny Greens Organic Farm of Urbana, Illinois, announced a recall of specific lots of Alfalfa Sprouts and Spicy Sprouts, because they were likely to be contaminated with *Salmonella*. The products were distributed in Illinois, Indiana and Missouri and they would have ended up in restaurants, groceries and supermarkets near those areas (CDC, 2010c).

Multistate Outbreak of *S.* Panama Infections Linked to Cantaloupe, 2011

In 2011, CDC in collaboration with the U.S. Food and Drug Administration (FDA) and with public health officials in a number of states, including California, Maryland, Oregon, and Washington, investigated and identified the likely source of a multistate outbreak caused by the pathogen *S.* Panama. As of 20th June, 2011, a total of 20 people from Arizona (1), California (2), Colorado (1), Maryland (1), Montana (1), Nevada (1), Oregon (6), Pennsylvania (1), Utah (1) and Washington (5) were reported to be infected with the outbreak strain. Ill people's age ranged from <1 year old to 68 years old, with the median age of 13 years old. Sixty-five percent were male. Three of the patients were hospitalized and no deaths were reported. Information gathered from patients connected the outbreak with the consumption of cantaloupes. Twelve of 16 ill people reported that they had eaten cantaloupe one week before illness and 11 of these 12 people had consumed cantaloupes purchased from

8 different locations of a national warehouse club. Investigation and produce trace back led to the conclusion that ill people had purchased cantaloupes harvested from a single farm in Guatemala. The last case of this outbreak was reported on 22nd April, 2011 (CDC, 2011a).

Multistate Outbreak of Human *S.* Enteritidis Infections Linked to Alfalfa Sprouts and Spicy Sprouts, 2011

From 12th April until 5th July, 2011, a total of 25 people infected with the strain of *S.* Enteritidis were reported. Ill people were from 5 states: Idaho (3), Montana (10), New Jersey (1), North Dakota (1) and Washington (10). According to patients' available information, illness began between 12th April and 15th June, 2011. Patients' age ranged from 12 years old to 77 years old and the median age was 35 years old. Seventy-six percent of them were female. From the 10 patients with the available information, 3 (30%) were hospitalized and no deaths were reported.

Collaborative investigations between local, state and federal public health and regulatory agencies linked the outbreak to Evergreen Fresh Sprouts, LLC Alfalfa Sprouts and Spicy Sprouts. On 1st July, 2011, Evergreen Fresh Sprouts, LLC of Moyie Springs, Idaho announced a recall of specific lots of alfalfa sprouts and spicy sprouts because these products had the potential to be contaminated with *Salmonella* spp. The products were distributed in Washington and Idaho by direct delivery to four distributors and three retail stores and could have ended up in restaurants and supermarkets in those areas and neighboring states (CDC, 2011b).

Multistate Outbreak of Human S. Agona Infections Linked to Whole, Fresh Imported Papayas, 2011

In the time period between 1st January and 25th August, 2011, 106 people were reported to be infected by *S.* Agona. The patients were identified in 25 states: Arkansas (1), Arizona (4), California (8), Colorado (1), Georgia (8), Illinois (18), Indiana (1), Kentucky (1), Louisiana (2), Massachusetts (1), Minnesota (3), Missouri (3), Nebraska (2), Nevada (1), New Jersey (1), New Mexico (3), New York (9), Ohio (1), Oklahoma (1), Pennsylvania (2), Tennessee (1), Texas (25), Virginia (2), Washington (5), and Wisconsin (2). Among the persons with available clinical information, illness began on or after 17th January, 2011. Ill people's age ranged between <1 year old and 91 years old, with the median age of 21 years old. Fifty-six percent were female and 11 people stated having travelled to Mexico one week before they became ill. Ten patients were hospitalized and no deaths were reported.

The outbreak was connected to the consumption of whole papayas. From the 56 ill people for whom information was available, 57% had consumed papayas one week before the illness onset. Product information reported from the patients showed that the company "Agromod Produce, Inc." of McAllen, Texas, was the common supplier of papayas purchased by the ill people. On 23rd July, 2011, Agromod Produce, Inc. voluntarily recalled fresh whole papayas because they had the potential to be contaminated with *Salmonella* spp. The products were imported from Mexico and distributed nationwide and also to Canada through retail stores and wholesalers. The recall included all Blondie, Yaya, Mañanita, and Tastylicious Brand papayas sold before the 23rd July, 2011.

On 25th August, 2011, an FDA Import Alert was issued denying the admission of papayas from Mexico into the United States, unless the importer proved that the products were not contaminated with *Salmonella*. The reason for this action was the fact that from 12th May, 2011 until 18th August, 2011, FDA analysis found 15.6% *Salmonella* contamination rate in papaya samples. The positive samples were from 28 different firms and included nearly all the major papaya producing regions in Mexico (CDC, 2011c).

Multistate Outbreak of *S.* Typhimurium and *S.* Newport Infections Linked to Cantaloupe, 2012

In 2012, a multistate outbreak of *S.* Typhimurium and *S.* Newport infected 261 people from 24 states: Alabama (25), Arkansas (6), Florida (1), Georgia (13), Illinois (36), Indiana (30), Iowa (9), Kentucky (66), Maryland (1), Michigan (8), Minnesota (2), Mississippi (7), Missouri (17), Montana (1), New Jersey (2), North Carolina (5), Ohio (5), Oklahoma (1), Pennsylvania (2), South Carolina (4), Tennessee (8), Texas (2), Virginia (1), and Wisconsin (9).

Two hundred twenty eight people were infected by *S.* Typhimurium and 33 by *S.* Newport. Clinical information was available for the 257 patients, according to which onset dates of the illness ranged from 6th July to 16th September, 2012. Fifty-five percent of the ill people were female. Patients' age ranged from <1 year old to 100 years old, with the median age of 47 years old. Ninety-four patients were hospitalized and three deaths were reported in Kentucky.

The investigation of the outbreak linked the multistate infections to cantaloupes originating from the company "Chamberlain Farms Produce, Inc." of Owensville, Indiana.

The strain of *S.* Typhimurium was isolated and thought to be responsible for the outbreak at first. The initial interviews of the first 24 patients reported that 18 (75%) of them had consumed cantaloupe during the week before their illness began. By the end of the outbreak 81 (65%) of the 123 patients interviewed reported having consumed cantaloupe one week before their illness.

Kentucky Division of Laboratory Services isolated the strain of *S.* Typhimurium from two cantaloupes collected from a retail location in Kentucky. Trace back investigations indicated that these cantaloupes originated from Chamberlain Farms. During August 2012, multistate outbreak of 33 *S.* Newport infections appeared. Investigation occurred by CDC, U.S. FDA and public health officials in several states suggested a possible connection between the *S.* Newport outbreak and the *S.* Typhimurium outbreak. According to the geographic distribution, dates of illness onset, and initial interview information the outbreak was linked to Chamberlain farms. Furthermore 13 (61%) of the 21 ill people reported eating cantaloupe during the week before illness.

On 22nd August, 2012, Chamberlain Farms, Inc. voluntarily recalled cantaloupes grown on its farm. Records available indicated that the product was initially shipped to Indiana, Kentucky, Missouri, Tennessee, Ohio, Illinois, and Wisconsin, although further shipment was likely (CDC, 2012a).

Multistate Outbreak of *S.* Braenderup Infections Linked to Mangoes, 2012

CDC in collaboration with the public health officials in several states and the U.S. FDA investigated a multistate outbreak of *S.* Braenderup infections linked to mangoes originating from a company in Mexico. A total of 127 people were reported to be infected with the outbreak strain. The reported states were: California (99), Delaware (1), Hawaii (4), Idaho (1), Illinois (2), Maine (1), Michigan (1), Montana (1), Nebraska (1), New Jersey (1), New York (3), Oregon (1), Texas (2), Washington (8), and Wisconsin (1). Clinical information were collected by 101 people from whom 33 (33%) reported being hospitalized. The illness onset dates ranged from 3rd July to 1st September, 2012, and no deaths were reported by the end of the outbreak. Patients ranged in age from <1 year old to 86 years old, with the median age of 33 years old and 56% of them were female. Fifty-one (63%) of 81 infected people reported having consumed mangoes one week before becoming ill.

California Department of Public Health (CDPH) and FDA conducted trace back investigations and identified the common source of contamination. Trace back analysis determined that mangoes purchased by ill people originated from "Agricola Daniella" a mango supplier with multiple plantations and a single packing house located in Sinaloa, Mexico. FDA conducted tests and isolated *Salmonella* from four mangoes originated from warehouse locations in the United States. These mangoes originated from Agricola Daniella. During July and August, 2012, a similar strain of *S.* Braenderup caused 21 illnesses in Canada. These illnesses were linked to mangoes and this fact lead to a recall of specific Daniella brand mangoes. On 29th August, 2012, Splendid Products of Burlingame, California, voluntarily recalled certain lots of Daniella brand mangoes and on 13th September, 2012, FDA placed Agricola Daniella on import alert, denying admission into the United States unless the importer proved that the products were not contaminated with *Salmonella*.

During August 2012, a multistate outbreak of 16 *S.* Worthington infections from 3 states: California (12), New Mexico (1), and Washington (3) occurred. Eight of the nine patients infected by *S.* Worthington reported that they had consumed mangoes during the week before their illness began. Furthermore, the fact that ill people were reported from similar states and during the same time period as *S.* Braenderup outbreak appeared, suggested a possible connection between the two outbreaks. Finally, one ill person in the *S.* Braenderup outbreak was also infected with *S.* Worthington. Nevertheless, no trace back investigations were conducted to determine the source of mangoes for these ill people. Onset dates of *S.* Worthington ranged between 19th July, 2012, and 12th September, 2012. Patients' age ranged from <1 year old to 86 years old and the median age was 65 years old. Sixty-three percent of the patients were female and among the 13 patients with available information, 3 (23%) were hospitalized. No deaths were reported (CDC, 2012b).

Multistate Outbreak of *Salmonella* Saintpaul Infections Linked to Imported Cucumbers, 2013

In April 2013, CDC in collaboration with the public health officials in many states and U.S. FDA started investigating an outbreak caused by *Salmonella* Saintpaul. Eighty-four people from 18 states were infected by the outbreak strain. The states with the reported ill people were: Arizona (11), California (29), Colorado (2), Idaho (2), Illinois (3), Louisiana (1),

Maryland (1), Massachusetts (1), Minnesota (9), Nevada (1), New Mexico (2), North Carolina (3), Ohio (3), Oregon (2), South Dakota (2), Texas (7), Virginia (3), and Wisconsin (2). According to the people with available clinical information, illness onset dates were from 12th January, 2013,until 28th April, 2013.Sixty-two percent of the ill people were female and their ages ranged from <1 year old to 89 years old, with the median age of 27 years old. Seventeen (28%) from the 60 patients with available information were hospitalized and no deaths were reported. Forty- nine ill people were interviewed and 34 (69%) of them reported having consumed various types of cucumbers purchased or consumed at multiple locations or restaurants. FDA traced cucumbers consumed by 6 ill people to the distributer "Tricar Sales, Inc." and to the suppliers "Daniel Cardenas Izabal" and "Miracle Greenhouse". On 24th April, 2013, the two suppliers were placed on import alert and cucumbers from these firms were denied admission in the United States unless the suppliers showed that the products were not contaminated with *Salmonella* (CDC, 2013).

MAJOR OUTBREAKS OF *SALMONELLA* SPP. IN EUROPE

Outbreak of *S.* Thompson Linked to Rucola Salad, 2004

In 2004 Nygard et al., reported an outbreak caused by *S.* Thompson. The outbreak was connected to the consumption of rucola lettuce (*Eruca sativa*, also known as rocket salad or arugula) and mixed salad in Norway. By 31st December, 2004, 21 cases had been reported and the onset date of illness stated to be on 24th October, 2004 (Nygard et al., 2008).On 25th November, 2004, Norwegian Food Safety Authority temporarily withdrew the implicated products from the market and on 26th November, 2004, the Swedish National Food Administration sent an alert through the EU Rapid Alert System for Food and Feed about finding *S.* Thompson in rucola lettuce (Nygard et al., 2004; 2008). The products in both Norway and Sweden came from the same producer in Italy. Furthermore, *Salmonella* serovars and *Cambypobacter* were also isolated from Italian produced rucola, in several countries. In order to control the outbreak, the Norwegian Food Safety Authority and the distributors of the implicated salad implemented measures that led to the lift of the ban on 14th December, 2004 (Nygard et al., 2004). At the same time period of the outbreak Sweden, England and Wales reported possible increases of the cases of *S.* Thompson since August 2004.

Despite the fact that only a few countries reported cases related to the outbreak of *S.* Thompson infections, the outbreak was thought to reflect a larger international outbreak caused by rucola, because a lot of different *Salmonella* serovars were detected in different countries. *Salmonella* was detected in rucola that was imported from Italy in Denmark (not specified type) and in Slovenia (*S.* Napoli) while in Norway *S.* Braenderup was isolated from a sample with mixed salad containing rucola (Nygard et al., 2004; 2008).

Oubreak of Salmonellosis Linked to Mug Beans, 2006

On 18th October, 2006, Environment and Health Administration (EHA) in Stockholm City informed the Department of Communicable Disease Control and Prevention (DCDC) in

Stockholm County about a suspected food poisoning at an Indian style restaurant located in the center of Stockholm. Initially 12 people contacted the restaurant and claimed suffering from food poisoning after having visited the restaurant. The manager of the restaurant informed the EHA after questioning the customers about the consumed menu and concluding that mug beans were the only common ingredient in the dishes of the patients. Finally, a total of 97 people contacted directly the restaurant to complain about food poisoning.

The mung beans, served in the restaurant, were bought from a wholesale trader that had imported them from Canada. Beans were put in lukewarm water and left over for 24 hours in room temperature and then they were rinsed and served. Considering that the mung beans were the source of the infection, the restaurant manager excluded them from the menu after 17th October, 2006. After omitting the beans from the menu, no cases were reported, a fact that strongly suggested that mug beans were the vehicle for the food borne disease.

Laboratory tests indicated that patients who ate at the restaurant between the end of September and 4th October were infected by *S.* Bareilly, whereas those who ate at the restaurant from 6th October until 16th October were infected by *S.* Virchow. Interestingly, none of them displayed both of the serotypes.

By the end of the outbreak, a total of 115 cases of salmonellosis were notified. Eighty-six of the patients were women and 29 were men. Ages of the patients ranged from 12 years old to 67 years old and the median ages were 34 and 35 years old for the females and males respectively. Thirteen people were hospitalized and 7 of them developed sepsis but no deaths were occurred. The number of the people affected by this outbreak was possible to have been higher considering the facts that restaurant served at about 300 portions each day and that only 72 of the 97 people who had initially contacted the restaurant to complain about food poisoning were found in the official reports (De Jong et al., 2007).

Outbreak of *S.* Seftenberg Linked to fresh Basil, 2007

In 2007, the Health Protection Agency (HPA) Laboratory of Enteric Pathogens (LEP) reported 51 human infections of *S.* Senftenberg in England and Wales. The reports started in the beginning of the year and were significantly higher than those of the previous years that were less than 10 in the same time period in 2005 and 2006. Onset date of illnesses was available for 23 patients ranging from 5th March to 12th May, 2007. From the 30 primary cases 60% were female and most of the patients were adults. Twenty of the 30 cases were interviewed and 30% of them reported having consumed fresh herbs three days before the onset of the illness, while 40% reported consumption of bagged salad.

As a part of a wide survey of fresh herbs on retail sale in the United Kingdom, samples of fresh basil were positive for the presence of *S.* Senftenberg. Seven samples of pre-packed fresh basil, grown in Israel, were also found to be contaminated with the outbreak strain. Additionally, after the complaint of a supermarket's customer in the Shetland Islands, fresh basil was tested and found to be contaminated with the outbreak strain of *S.* Senftenberg.

In April, 2007, 3 cases were reported from Scotland, and all of them matched with the outbreak profile according to the PFGE profiles. One of these patients stated having travelled to Tenerife. An increase in cases of *S.* Senftenberg was also reported from Denmark in 2007; a total of 11 cases were reported whilst none cases had been reported in the first six months of 2006. Three of the cases matched the outbreak strains' PFGE profiles from the U.K. Two of them reported that their exposure probably occurred during their visit in the U.S.A., Poland or the U.K. Five cases of *S.* Senftenberg infection were also reported in the Netherlands in January (2) and in May, 2007 (3). Two of the cases matched with the outbreak profile and one patient, who was diagnosed in May, reported having consumed a mixed pasta salad, possibly mixed with fresh basil one week before the symptoms started. In addition, the United States and Israel reported human isolates of *S.* Senftenberg.

On 25th May, 2007, the U.K. Food Standards Agency (FSA) issued a public statement, warning the consumers about the contaminated batches of the basil. In parallel a withdrawal of products took place in three U.K. supermarket chains. Specifically, ASDA, Sainsbury's and Somerfield stores withdrawal all the potentially affected products from their shelves. A notice was also issued on the Alert System for Food and Feed (RASFF) by the European Commission on 25thMay, 2007. In 2007, Pezzoli et al. reported that their investigations confirmed that the outbreak strain of *S.* Senftenberg had been identified in fresh basil grown in Israel, in the U.K, in Scotland, Denmark, the Netherlands and the United States of America as well as in patients who reported travelling to Tenerife and Morocco.

Outbreak of *S.* Weltevreden Linked to Alfalfa Sprouts, 2007

Between 10th and 15th October, 2007, the Norwegian Institute of Public Health (FHI) isolated *S.* Weltevreden from four patients with gastroenteritis. Patients were living in south-eastern Norway and reported not travelling abroad for the time period of one month before the onset of the illness. On 19th October, 2007, FHI in collaboration with the municipal medical officers and the Norwegian Food Safety Authority (NFSA) started investigating the outbreak in order to identify the source of the outbreak. On 22nd October, 2007, an urgent notice was sent out through the European Centre for Disease Prevention and Control (ECDC) and Denmark responded that at that time 18 cases of *S.* Weltevreden were under investigation. The onset date of illness was in the end of July. On 26th October, Finland reported 7 cases that had occurred between 1stAugust and 1stOctober, 2007.

Based on the investigations and the available information, it was concluded that alfalfa sprouts grown from contaminated seeds were the source of the outbreak in all three countries.On 23rd October, 2007, *S.* Weltevreden was isolated by a major Danish alfalfa sprout producer and on the same day the Danish authorities issued a warning through the Rapid Alert System for Food and Feed (RASFF). In addition, *S.* Weltevreden was also isolated by sprouts sold in Finland.

Seeds used for the grown of alfalfa sprouts were probably Italian and they had been imported to Denmark in July and August 2007. On 19th September, 2007, the Danish producer exported part of the batch of the seeds to a Norwegian alfalfa sprout producer. According to invoices, the batch of seeds used in Denmark and Norway was traded via retailers in Netherlands and Germany to Denmark. As far as Finland was concerned, except for the fact that seeds used there came from a Dutch supplier no further connections were found, since the batch of the sprouts was imported to Finland in June but it was not on the market before August. On 18th, 23rd and 28th October, 2007, the alfalfa sprouts were recalled and withdrawn in Denmark, Norway and Finland, respectively.

Until 19thNovember, 19 cases had been reported in Norway, 19 in Denmark, and 7 in Finland. The onset dates of illnesses for the 17 Norwegian patients, for who clinical information was available, ranged from 28th September to 15th October. For the 14 Danish patients the onset dates were between 23rd July and 20th October and for the 5 Finnish patients the onset dates of illness were between 11th August and 30th September.The patients' ages ranged from 18 to 83 years old with the median age of 34 years old (Emberland et al., 2007).

Outbreak of *S.* Newport Linked to Mung Sprouts, 2011

The largest *S.* Newport outbreak reported in Germany occurred in October and November, 2011. At the same time period cases related to the outbreak *S.* Newport were reported in the Netherlands.

Mung beans were suspected as the source of the outbreak because *S. Newport* had been isolated from mung beans of a sprout distributor in Germany on 19th October, 2011, during a routine sampling and also because an unspecified serovar of *Salmonella* had been detected in a sprout producer A in the Netherlands on 18th and 21st October, 2001, during a sampling of the company in its own production batches. Trace back of the products showed that the mung bean sprouts originated from one of the lots, were produced in the Netherlands. The distributor had received the sprouts from the producer A on 18th October, 2011. The above findings were notified to the European food safety and public health authorities through the Rapid Alert System for Food and Feed (RASFF) of the European Commission on 17th November, 2011.

One hundred six infections in Germany and 20 infections in the Netherlands were finally associated with this outbreak that was caused by a single strain of *S.* Newport. The onset of illness in Germany was between 20th October and 8th November, 2011, while in the Netherlands was between 13th October and 1st November, according to 15 of the Dutch patients for whom information was available.

In Germany, patients' age ranged from <1 year old to 91 years old and the median age was 38 years old. Fifty-two percent of the patients were female. Twenty-eight percent of the cases were reported to be hospitalized but no fatal cases were reported. In the Netherlands patient's age ranged from 1o years old to 89 years old, with the median age of 37 years old. Fifteen of the patients were female. Fifteen people were hospitalized, 2 were under hospitalization when their infection occurred and for the remaining 3 patients no clinical information was available. Fifteen of the 20 Dutch patients were interviewed about the outbreak. Four of them reported having or possibly having consumed sprouts during the week before the onset of the illness, 9 reported having eaten meals in which sprouts are usually used or could be used and for the 2 remaining no connection to sprout consumption was found.

A distributor of the rehabilitation clinic in Northern Germany traced back his sprouts to the producer A in the Netherlands. In addition, six Asian restaurants, visited by patients before their illness, had received mung sprouts from the sprout producer A, through several distributors. In the Netherlands, sprouts served in the affected hospital, were traced back via several distributors to producer A. The restaurants reported that served sprouts were either briefly heated (when sprouts were added in the dish shortly before serving it) or well cooked (when spouts were cooked with the dish). On the contrary, the sprouts served at the salad bar of the rehabilitation clinic in the Northern Germany were uncooked. By the end of the investigation, the implicated lots of the mung bean sprouts had already expired so it was considered that the products had already been consumed or discarded. In addition, the implicated seed lots of the producer A that were positive for *Salmonella* were blocked from the market (Bayer et al., 2014).

SALMONELLA SEROVARS ON HUMANS

In 2012, European Food Safety Authority (EFSA) and European Centre for Disease Prevention and Control (ECDC) reported the distribution of the 10 most common *Salmonella* serovars in humans, in an E.U. summary on zoonoses, zoonotic agents and food-borne outbreaks during 2010. Data sourced from 96,745 cases in the 26 E.U. member states: Austria, Belgium, Cyprus, Czech Republic, Denmark, Estonia, Finland, France, Germany, Greece, Hungary, Ireland, Italy, Latvia, Lithuania, Luxembourg, Malta, Netherlands, Poland, Portugal, Romania, Slovakia, Slovenia, Spain, Sweden and United Kingdom. The two most frequently reported serovars in humans during 2010 were *S.* Enteritidis (45%) and *S.* Typhimurium (22.4%). *S.* Infantis was the third most commonly reported serovar (1.8%) followed by *S.* Typhimurium, monophasic1,4,[5],12:i:- (1.5%), *S.* Newport (0.9%), *S.* Kentucky (0.8%), *S.* Virchow (0.7%), *S.* Derby (0.7%), *S.* Mbandaka (0.5%), *S.* Agona (0.5%) and other serovars (25.3%).

Based on the major outbreaks that have been reported from 1989 until today and have been linked to the consumption of fresh fruits and vegetables, the two most commonly reported *Salmonella* serovars are *S.* Saintpaul (31.2%) and *S.* Typhimurium (18.6%). *S.* Poona (7.1%), *S.* Montevideo (6.3%) and *S.* Javiana (3.8%) were also frequently reported, as well as a lot of other serovars, such as *S.* Newport (3.6%), *S.* Enteritidis (1.6%) and serovar 1,4,[5],12:i:-(Figure 1).

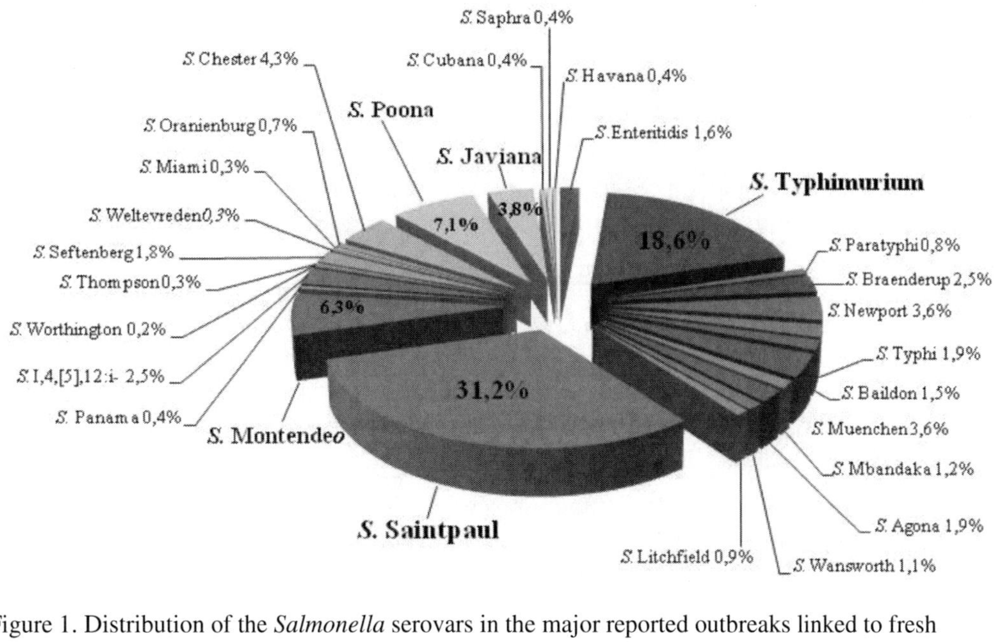

Figure 1. Distribution of the *Salmonella* serovars in the major reported outbreaks linked to fresh products, (1898-2013).

ACKNOWLEDGMENTS

The authors gratefully acknowledge funding from the European Union Seventh Framework Programme (FP7/2007-2013) under grant agreement n. 289719 (Project QUAFETY).

REFERENCES

Al-Hindawi, N; Rished, R. Presence and distribution of *Salmonella* species in some local foods from Baghdad City, Iraq. *J. Food Prot.*, 1979, 42(11), 877-880.

Backer, HD; Mohle-Boetani, JC, Werner SB; Abbott, SL; Farrar, J; Vugia, DJ. High incidence of extra-intestinal infections in a *Salmonella* Havana outbreak associated with alfalfa sprouts. *Pub. Health Rep.*, 2000, 115, 339-345.

Bayer, C; Bernard, H; Prager, R; Rabsch, W; Hiller, P; Malorny, B; Pfefferkorn, B; Frank, C; de Jong, A; Friesema, I; Stark, K; Rosner, BM. An outbreak of *S.* Newport associated with mung bean sprouts in Germany and the Netherlands, October to November 2011. *EuroSurveill.*, 2014, 19(1):pii=20665. http://www.eurosurveillance.org/ViewArticle.aspx?ArticleId=20665.

Beuchat, LR. Pathogenic microorganisms associated with fresh produce. *J. Food Prot.*, 1996, 59(2), 204-16.

Birkhead, GS; Morse, DL; Levine, WC; Fudala, JK; Kondracki, SF; Chang, HG; Shayegani, M; Novick, L; Blake, PA. Typhoid fever at a resort hotel in New York: a large outbreak with an unusual vehicle. *J. Infect. Dis.*, 1993, 167, 1228-1232.

Blostein, J. An outbreak of *S.* javiana associated with consumption of watermelon. *J. Environ.Health.*1993, 56(1), 29-31.

Butler, ME. *Salmonella* outbreak leads to juice recall in Western states. *Food Chem. News,* 2000, 42(10), 19.

Centers of Disease Control and Prevention (CDC).*Salmonella* Typhimurium outbreak traced to a commercial apple cider--New Jersey.1975, MMWR 24:87.

Centers of Disease Control and Prevention (CDC). *Salmonella* Oranienburg gastroenteritis associated with consumption of precut watermelons-Illinois. 1979, MMWR 28, 522-523.

Centers of Disease Control and Prevention (CDC). Multi-state outbreak of *Salmonella* Poona infections - United States and Canada 1991. 1991, MMWR 40, 549-552.

Centers of Disease Control and Prevention (CDC). Outbreak of *S.* Hartford among travellers to Orlando, FA, May 1995. 1995, Atlanta, GA: CDC.

Centers of Disease Control and Prevention (CDC). Outbreak of *Salmonella* serotype Muenchen infections associated with unpasteurized orange juice--United States and Canada, July 1999. 1999, MMWR 48, 582-585.

Centers of Disease Control and Prevention (CDC). Multistate outbreak of *S.* Typhimurium infections linked to tomatoes (final update). 2006, http://www.cdc.gov/salmonella/typh2006/index.html. Page last updated: 13 May, 2008.

Centers of Disease Control and Prevention (CDC). Multistate outbreak of *S.* Wandsworth infections linked to Veggie Booty (final update). 2007, http://www.cdc.gov/salmonella/wandsworth/index.html.Page last updated: 13 May, 2008.

Centers of Disease Control and Prevention (CDC). Multistate outbreak of *S.* Litchfield infections linked to cantaloupe (final update). 2008a, http://www.cdc.gov/salmonella/litchfield/index.html. Page last updated: May 13, 2008.

Centers of Disease Control and Prevention (CDC). Multistate outbreak of *S.* Saintpaul infections linked to raw produce (final update). 2008b, http://www.cdc.gov/salmonella/saintpaul-jalapeno/index.html. Page last updated: August 25, 2011.

Centers of Disease Control and Prevention (CDC). Multistate outbreak of *Salmonella* infections linked to pistachio nuts (Final Update). 2009a, http://www.cdc.gov/salmonella/pistachios/index.html. Page last updated: August 25, 2011.

Centers of Disease Control and Prevention (CDC). Multistate outbreak of human *S.* Saintpaul infections linked to raw alfalfa sprouts (final update). 2009b, http://www.cdc.gov/salmonella/saintpaul-alfalfa/index.html.

Page last updated: August 25, 2011.

Centers of Disease Control and Prevention (CDC). Multistate outbreak of human *S.* Montevideo infections (final update). 2010a, http://www.cdc.gov/salmonella/montevideo/index.html. Page last updated: December 2, 2012.

Centers of Disease Control and Prevention (CDC). Multistate outbreak of human *S.* Newport infections linked to raw alfalfa sprouts (final update). 2010b, http://www.cdc.gov/salmonella/newport/index.html. Page last updated: June 29, 2010.

Centers of Disease Control and Prevention (CDC). Multistate outbreak of human *Salmonella* I 4, [5], 12:i:- infections linked to alfalfa sprouts (final update). 2010c, http://www.cdc.gov/salmonella/i4512i-/index.html. Page last updated: August 25, 2011.

Centers of Disease Control and Prevention (CDC). What is Salmonellosis? 2010d, http://www.cdc.gov/salmonella/general/. Page last updated: April 5, 2012

Centers of Disease Control and Prevention (CDC). Multistate outbreak of *S.* Panama infections linked to cantaloupe. 2011a, http://www.cdc.gov/salmonella/panama0311/062311/index.html. Page last updated: August 25, 2011.

Centers of Disease Control and Prevention (CDC). Multistate outbreak of human *S.* Enteritidis infections linked to alfalfa sprouts and spicy sprouts. 2011b, http://www.cdc.gov/salmonella/sprouts-enteritidis0611/070611/index.html. Page last updated: July 6, 2011.

Centers of Disease Control and Prevention (CDC). Multistate outbreak of human *S.* Agona infections linked to whole, fresh imported papayas. 2011c, http://www.cdc.gov/salmonella/agona-papayas/082911/index.html. Page last updated: August 29, 2011.

Centers of Disease Control and Prevention (CDC). Multistate outbreak of *S.* Typhimurium and *S.* Newport infections linked to cantaloupe (final update). 2012a, http://www.cdc.gov/salmonella/typhimurium-cantaloupe-08-12/index.html. Page last updated: October 9, 2012.

Centers of Disease Control and Prevention (CDC). Multistate outbreak of *S.* Braenderup infections associated with mangoes (final update). 2012b, http://www.cdc.gov/salmonella/braenderup-08-12/index.html. Page last updated: October 11, 2012.

Centers of Disease Control and Prevention (CDC). Multistate outbreak of *S.* Saintpaul infections linked to imported cucumbers (final update). 2013, http://www.cdc.gov/salmonella/saintpaul-04-13/index.html. Page last updated: June 20, 2013.

Collins, JE. Impact of changing consumer lifestyles on the emergence/re-emergence of foodborne pathogens. *Emerg. Infect. Dis.*, 1997, 3(4), 471–479.

Cook, KA; Dobbs, TE; Hlady, WG; Wells, JG; Barrett, TJ; Puhr, ND; Lancette, GA; Bodager, DW; Toth, BL; Genese, CA; Highsmith, K; Pilot, E; Finelli, L; Swerdlow, DL. Outbreak of *Salmonella* serotype Hartford infections associated with unpasteurized orange juice. *J. Am. Med. Assoc.*, 1998, 280, 1504-1509.

De jong, B; Öberg, J; Svenungsson, B. Outbreak of salmonellosis in a restaurant in Stockholm, Sweden, September – October 2006. *EuroSurveill.*, 2007, 12(11):pii=749. http://www.eurosurveillance.org/ViewArticle.aspx?ArticleId=749.

Deeks, S; Ellis, A; Ciebin, B; Khakhria, R; Naus, M; Hockin, J. *Salmonella* Oranienburg, Ontario.*Can. Comm. Dis. Rep.*, 1998, 24, 177-179

Duncan, TG;Doull, JA; Miller, ER; Bancroft, H. Outbreak of typhoid fever with orange juice as the vehicle, illustrating the value of immunization. *Am. J. Public Health*, 1946, 36, 34-36.

European Food Safety Authority (EFSA), European Centre for Disease Prevention and Control (ECDC).The European Union summary report on trends and sources of zoonoses, zoonotic agents and food-borne outbreaks in 2010. *EFSA Journal,* 2012, 10(3), 2597. http://www.efsa.europa.eu/en/efsajournal/doc/2597.pdf.

Emberland, KE; Ethelberg, S; Kuusi, M; Vold, L; Jensvoll, L; Lindstedt, BA; Nygård, K; Kjelsø, C; Torpdahl, M; Sørensen, G; Jensen, T; Lukinmaa, S; Niskanen, T; Kapperud, G. Outbreak of *S.* Weltevreden infections in Norway, Denmark and Finland associated with alfalfa sprouts, July-October 2007. *EuroSurveill.*, 2007, 12, 10–12.

Ercolani, GL. Bacteriological quality assessment of fresh marketed lettuce and fennel. *Appl. Environ. Microbiol.*, 1976, 31(6), 847-852.

Fatica, MK; Schneider KR. *Salmonella* and produce: Survival in the plant environment and implications in food safety. *Virulence,* 2011, 2, 573–579.

Feng, P. A. Summary of background information and foodborne illness associated with the consumption of sprouts. Food and Drug Administration, Center for Food Safety and Applied Nutrition, 1999.

Food and Drug Administration (FDA). FDA Survey of imported fresh produce FY 1999 Field Assignment. 2001, http://www.fda.gov/Food/GuidanceRegulation/GuidanceDocuments RegulatoryInformation/ProducePlantProducts/ucm118891.htm. Page Last Updated: 8 June 2014

Food and Drug Administration (FDA). Analysis and evaluation of preventive control measures for the control and reduction/elimination of microbial hazards on fresh and fresh-cut produce. Chapter IV: Outbreaks Tables. 2013, http://www.fda.gov/ Food/FoodScienceResearch/SafePracticesforFoodProcesses/ucm091270.htm. Page Last Updated: 6 February 2013.

Food and Drug Administration (FDA). Analysis and evaluation of preventive control measures for the control and reduction/elimination of microbial hazards on fresh and fresh-cut Produce. Chapter IV: Incidence Tables. 2013, http://www.fda.gov/Food/FoodScienceResearch/SafePracticesforFoodProcesses/ucm091 268.htm. Page Last Updated: 24 April 2013.

Garcia-Villanova Ruiz, B; Cueto Espinar, A;Bolanos Carmona, MJ. A comparative study of strains of salmonella isolated from irrigation waters, vegetables and human infections. *Epidemiol. Infect.,* 1987a, 98, 271-276.

Garcia-Villanova Ruiz, B; Galvez Vargas, R; Garcia-Villanova, R. Contamination on fresh vegetables during cultivation and marketing. *Int. J. Food Microbiol.*, 1987b, 4, 285-291.

Gayler, GE; MacCready, RA; Reardon, JP; McKernan, BF. An outbreak of Salmonellosis traced to watermelon. *Public Health Rep*., 1955, 70(3), 311-313.

Havelaar, AH; Ivarsson, S; Löfdahl, M; Nauta, MJ. Estimating the true incidence of campylobacteriosis and salmonellosis in the European Union, 2009. *Epidemiol Infect.,* 2012, 1-10.

Jain, S; Bidol, SA; Austin, JL; Berl, E; Elson, F; LeMaile-Williams, M; Deasy III, M; Moll, ME; Rea, V; Vojdani, JD; Yu, PA; Hoekstra, RM.; Braden, CR; Lynch, MF. Multistate Outbreak of *S.* Typhimurium and Saintpaul Infections Associated with Unpasteurized Orange Juice—United States, 2005. *Clin. Infect. Dis*., 2009, 48, 1065–1071.

Jerngklinchan, J; Saitanu, K. The occurrence of *Salmonellae* in bean sprouts in Thailand. *Southeast Asian J. Trop. Med. Public Health*, 1993, 24(1), 114-118.

Lemunier, M; Francou, C; Rousseaux, S; Houot, S; Dantigny, P; Piveteau, P; Guzzo, J. Long-Term survival of pathogenic and sanitation indicator bacteria in experimental biowaste composts. *Appl. Environ. Microbiol*., 2005, 71(10), 5779.

Lin, CM; Fernando, SY; Wei, CI. Occurrence of *Listeria monocytogenes, Salmonella* spp., *Escherichia coli* and *E. coli* O157:H7 in vegetable salads. *Food Control*, 1996, 7, 135-140.

Little, C; Roberts, D; Youngs, E; de Louvois, J. Microbiological quality of retail imported unprepared whole lettuces: a PHLS Food Working Group Study. *J. Food Prot*., 1999, 62(4), 325-328.

Lund, BM; Snowdon, AL. Fresh and processed fruits, Chapter 27. In: Lund BM, Baird-Parker TC, Gould GW, editors. The microbiological safety and quality of food, Volume I. Gaithersburg (MD): Aspen, 2000, pp. 738-758.

Madden, JM. Microbial pathogens in fresh produce-the regulatory perspective. *J. Food Prot.,* 1992, 55(10), 821-823.

Mahon, BE; Ponka, A; Hall, WN; Komatsu, K; Dietrich, SE; Siitonen, A; Cage, G; Lambert-Fair, MA; Hayes, PS; Bean, NH; Griffin, PM; Slutsker, L. An international outbreak of Salmonella infections caused by alfalfa sprouts grown from contaminated seeds. *J. Infect. Dis*., 1997, 175, 876-882.

Mohle-Boetani, JC; Reporter, R; Werner, SB; Abbott, S; Farrar, J; Waterman, SH; Vugia, DJ. An outbreak of *Salmonella* serogroup Saphra due to cantaloupes from Mexico.*J. Infect. Dis.,* 1999, 180, 1361-1364.

Nguyen-the, C; Carlin, F. Fresh and processed vegetables. In: Lund BM, Baird-Parker TC, Gould GW, editors. The microbiological safety and quality of food. Gaithersburg (MD): Aspen, 2000, pp. 620-684.

Nygård, K; Lassen, J; Vold, L; Aavitsland, P; Fisher, IS. International outbreak of *S.* Thompson caused by contaminated ruccola salad - update. *Euro Surveill.* 2004, 8(51): pii=2602. http://www.eurosurveillance.org/ViewArticle.aspx?ArticleId=2602.

Nygård, K; Lassen, J; Vold, L; Andersson, Y; Fisher, I; Löfdahl, S; Threlfall, J; Luzzi, I; Peters, T; Hampton, M; Torpdahl, M; Kapperud, G; Aavitsland, P. Outbreak of S. Thompson infections linked to imported rucola lettuce. *Foodborne Pathog. Dis.,* 2008, 5(2), 165-173.

Pao, S; Brown, GE; Schneider, KR. Challenge studies with selected pathogenic bacteria on freshly peeled hamlin orange. *J. Food Sci.,* 1998, 63(2), 359-362.

Paquet, PE. Epidemie de fievretyphoide: Determinee par la consommation de petit cidre. *Revue d' Hygiene*, 1923, 45, 165-169.

Pezzoli, L; Elson, R; Little, C; Fisher, IS; Yip, H; Peters, TM; Hampton, M; de Pinna, E; Coia, JE; Mather, H; Brown, DJ; Møller Nielsen, E; Ethelberg, S; Heck, ME; de Jager, CM; Threlfall, J. International outbreak of *S.* Senftenberg in 2007. *Euro Surveill.*, 2007, 12(24):pii=3218.

Ponka, A; Andersson, Y; Siitonen, A; de Jong, B; Johhola, M; Halkala, O; Kuhmonan, A; Pakkala, P. *Salmonella* in alfalfa sprouts. *Lancet,* 1995, 345, 462-463.

Rude, RA; Jackson, GJ; Bier, JW; Sawyer, TK; Risty, NG.Survey of fresh vegetables for nematodes, amoebae, and *Salmonella. J. AOAC Int.*, 1984, 67, 613-615.

Saddik, MF; El-Sherbeeny, MR; Bryan, FL. Microbiological profiles of Egyptian raw vegetables and salads. *J. Food Prot.,* 1985, 48(10), 883-886.

Scallan, E; Hoekstra, R; Angulo, F; Tauxe, R; Widdowson, M; Roy, S; Jones, J; Griffin, P. Foodborne illness acquired in the United States—major pathogens. *Emerg. Infect. Dis.,* 2011, 17, 7-15.

Splittstoesser, DF;Queale, DT;Andaloro, BW. The microbiology of vegetable sprouts during commercial production. *J. Food Safety*, 1983, 5, 79-86.

Susman, E. Nationwide outbreak of Salmonella linked to tomatoes: experts call for stricter measures to decrease spread of disease causing bacteria. WebMD Health. 1999. http://my.webmd.com/content/article/1728.50099.

Tamminga, SK; Beumer, RR; Kampelmacher, EH. The hygenic quality of vegetables grown in or imported into the Netherlands: a tentative study. *J. Hyg. Camb.*, 1978, 80, 143-154.

Taormina, PJ; Beuchat, LM; Slutsker, LM. Infections associated with eating seed sprouts: an international concern. *Emerg. Infect. Dis.*, 1999, 62, 626-634. http://www .eurosurveillance.org/ViewArticle.aspx?ArticleId=3218.

Tauxe, RV. Emerging foodborne diseases: an evolving public health challenge. *Dairy, Food Environ. Sanit.*, 1997, 17(12), 788-795.

Tauxe, R; Kruse, H; Hedberg, C; Potter, M; Madden, J; Wachsmuth, K. Microbial hazards and emerging issues associated with produce—A preliminary report to the National Advisory Committee on Microbiologic Criteria for Foods. *J. Food Prot.,* 1997, 60, 1400-1408.

Tournas VH. Spoilage of vegetable crops by bacteria and fungi and related health hazards. *Crit. Rev. Microbiol.,* 2005, 31, 33-44.

Wei, CI; Huang, TS; Kim, JM; Lin, WF; Tamplin, ML; Bartz, JA. Growth and survival of *Salmonella* Montevideo on tomatoes and disinfection with chlorinated water. *J. Food Prot.*, 1995, 58(8), 829-836.

World Health Organization (WHO).Salmonella (non-typhoidal).Fact sheet N°139.2013, http://www.who.int/mediacentre/factsheets/fs139/en/. Page last updated August 2013.

In: Salmonella
Editor: Christopher B. Hackett

ISBN: 978-1-63463-651-3

Chapter 6

ATTACHMENT OF *SALMONELLA* SPP. TO FOOD CONTACT AND PRODUCT SURFACES AND BIOFILM FORMATION ON THEM AS STRESS ADAPTATION AND SURVIVAL STRATEGIES

Efstathios Giaouris[1] and Live L. Nesse[2]
[1]Department of Food Science and Nutrition, Faculty of the Environment, University of the Aegean, Myrina, Lemnos Island, Greece
[2]Section for Bacteriology-Aquatic and Terrestrial, Norwegian Veterinary Institute, Oslo, Norway

ABSTRACT

The ability of various microorganisms to attach to surfaces and create biofilms on them is rather a cause of concern for many industries, including for those occupied with food production and processing. Thus, the attachment of bacterial pathogens to food processing equipment is considered as an essential contributing factor in foodborne disease outbreaks, since this may ultimately lead to the contamination of food products. Improperly cleaned surfaces promote soil build-up, and, in the presence of water, contribute to the development of microbial biofilms which may contain pathogenic bacteria, such as *Salmonella*. It is well recognized that biofilm cells differ physiologically from their planktonic counterparts, presenting a modified and heterogeneous gene expression profile. Additionally, it has been observed that the resistance of sessile cells to antimicrobials and other environmental stresses is significantly increased compared to what is normally seen with the same cells being planktonic. Noteworthily, salmonellae have been shown to survive for years in non-enteric habitats, including sessile communities on food contact and product surfaces. Indeed, several reports have demonstrated the ability of *Salmonella* to attach and form biofilms on abiotic surfaces, such as stainless steel, plastic, rubber, glass, marble and cement. *Salmonella* is also able to strongly attach and persist on both animal and plant (produce) surfaces. It is believed that the attachment to all these surfaces and the subsequent biofilm formation on them enhance the capacity of pathogenic *Salmonella* bacteria to successfully cope with hurdles that are commonly encountered outside the host and within food processing. The purpose of this chapter is to review the current available knowledge related to the attachment of *Salmonella* to food contact and product surfaces and the possible subsequent sessile development on them in view of the strong impact of these two interrelated capabilities on the enhancement of its survival outside the host, its environmental persistence and spread. Undoubtedly, the ability to recognize why and how *Salmonella* attach to such surfaces is an important area of focus, since this may provide valuable ways towards the elimination of this important pathogen from food processing environments and eventually lead to reduced *Salmonella*-associated human illness.

Keywords: *Salmonella enterica*, attachment, biofilm, food, surfaces, stress, adaptation, survival

1. Introduction

Salmonella enterica is one of the most significant enteric foodborne bacterial pathogens, with host-adapted strains able to cause systemic human infections and persist for long time periods, posing therefore significant public-health problems (Ruby et al., 2012). Non-typhoidal *Salmonella* strains are classified by serological characteristics into more than 2500 serovars of which the serovars Typhimurium and Enteritidis are the most prevalent (Foley et al., 2013). *S. enterica* encounter a variety of unfavorable environmental conditions, while passing between numerous natural, industrial and host environments, ranging from temperature fluctuations, nutrient availability and changes in osmolarity, to the presence of preservatives, antimicrobial peptides and reactive oxygen/nitrogen species (Runkel et al., 2013). Such fluctuating conditions impact on various areas of its cellular physiology, including virulence, growth and antimicrobial resistance. Amazingly, these bacteria have evolved multiple, complex, often interconnected systems of stress management, as part of their survival strategies, making them incredibly adept and versatile (Foster and Spector, 1995; Audia et al., 2001). The successfully control of all these overlapping stress response networks is achieved by a coordinated and complex programme of gene expression and protein activity, involving a range of transcriptional regulators, sigma factors and two component regulatory systems (Dorman, 2009). Integrating the various stress responses into a multi-defense strategy allows the organism to anticipate, survive and persist in the various non-host and host environments it encounters (Shen and Fang, 2012; Spector and Kenyon, 2012).

Besides *Salmonella*'s tremendous ability to sense, adapt and respond to a diverse range of adverse environmental conditions, these bacteria are well known to attach to various both abiotic and biotic surfaces, such as those of plants, the eukaryotic host, industrial facilities, and medical supplies, and create biofilms (Giaouris et al., 2012; Steenackers et al., 2012). These are assemblages of microorganisms adherent to each other and/or to a surface and embedded in a matrix of self-produced extracellular polymeric substances (EPS) (Stoodley et al., 2002; Hall-Stoodley et al., 2004). Interestingly, the adhesive properties of fimbriae salmonellae to animal, plant and fungal cells have been early described (Duguid et al., 1966). Last decades, biofilm formation by bacterial pathogens, such as *Salmonella*, has attracted much attention, mainly in the medical and food processing fields, due to its potential risks, including antimicrobial resistance and virulence factor production (Stewart and Costerton,

2001; Chmielewski and Frank, 2003). Thus, the persistence of *Salmonella* inside biofilms within the food chain constitutes a major health concern, as it is believed that biofilm formation enhances the capacity of pathogenic *Salmonella* bacteria to survive stresses that are commonly encountered within food processing (Vestby et al., 2009b). There is strong evidence of *Salmonella* survival in food processing environments, with these bacteria to be routinely isolated from filters, floors, tables, mixers, mincing machines and other processing equipment, even over long time periods (Nesse et al., 2003; Gounadaki et al., 2008). Indeed, it has been reported that once a production line is contaminated with *Salmonella*, the microorganism will establish itself on the equipment and facility surfaces (Margas et al., 2014).

While the molecular mechanisms required for the survival of bacteria on surfaces are not fully understood, transcriptional studies have demonstrated that biofilm growth triggers the expression of specific sets of genes (Puttamreddy et al., 2008). Thus, it has become clear that biofilm cells express properties distinct from planktonic cells, one of which is an increased resistance to antimicrobial agents (Anderson and O'Toole, 2008; Coenye, 2010). Recent research has begun to shed light on how and why surface-attached microbial communities develop such resistance, with several mechanisms presumably involved in this, including: (i) limitations to the free diffusion of antimicrobial agents through the biofilm matrix, (ii) variability in the physical and chemical microenvironments within the biofilm (e.g., varied conditions of pH, osmotic strength, or nutrients), leading to varied levels of metabolic activity within the biofilm milieu and also to alteration of sanitizer's efficiency, (iii) adaptive stress responses, resulting from mutations, stochastic gene expression and also through possible horizontal transfer of genes coding for resistance mechanisms (e.g. detoxifying membrane transporters), and (iv) the differentiation of bacterial cells into physiological states less susceptible to treatments (e.g. dormant, viable but not culturable, VBNC), together with the presence of extremely resistant "persister" cells (Mah and O'Toole, 2001; Gilbert et al., 2002a,b; Fux et al., 2005). Noteworthily, cells in shearable and nonshearable regions of *S.* Enteritidis biofilms have been found to be morphologically and physiologically distinct (Mangalappalli-Illathu et al., 2009). These stratified patterns of cell metabolism and morphology in biofilms may yield important information of how cells of deeply embedded biofilm bacteria survive imposed conditions of stress, such as treatment with antimicrobial agents or antibiotics. Importantly, owing to the heterogeneous nature of a biofilm, it is likely that multiple resistance mechanisms work in parallel within a single community.

Regarding *S. enterica*, there are a number of interesting studies occupied with its cellular physiology inside a biofilm (Kim and Wei, 2009; White et al., 2010; Steenackers et al., 2012). In this bacterium, the expression of the main extracellular matrix components stabilizing the biofilm structure is dependent on the transcriptional regulator CsgD, whose expression integrates many environmental signals, such as starvation, oxygen tension, temperature, pH and osmolarity (Gerstel et al., 2003; Gerstel and Römling, 2001, 2003; Römling, 2005). In most wild type *Salmonella* strains, this is maximum at temperatures below 30 °C, but mutations in the *csgD* promoter can lead to its expression independently of temperature (Romling et al., 1998). CsgD is the major regulator of the rdar colony morphotype which *S. enterica* biofilm forming strains present when these are cultured on Congo red agar plates, due to the red dry and rough colony appearance. Interestingly, White et al. (2010) showed by comparing extracellular matrix-embedded, wild-type *S.* Typhimurium and the matrix-deficient *csgD* mutant that the two populations present distinct metabolite and gene expression patterns, with wild-type cells expressing genes mainly involved in gluconeogenesis and stress-resistance pathways. CsgD positively regulates the transcription of the *csgBAC* operon, which encodes the structural subunits for curli fimbriae (Hammar et al., 1995), and contributes indirectly to cellulose production by activating the transcription of *adrA* (Gerstel and Römling, 2003). AdrA is a diguanylate cyclase that synthesizes the second messenger signaling molecule bis-(3'-5')-cyclic dimeric guanosine monophosphate (cyclic-di-GMP), the effector molecule that binds to and allosterically activates cellulose synthase (Simm et al., 2004). c-di-GMP is widespread throughout the bacterial domain and plays a vital role in regulating the transition between the motile planktonic lifestyle and the sessile biofilm forming state (Cotter and Stibitz, 2007; Ahmad et al., 2011; Le Guyon et al., 2014). GGDEF and EAL domain proteins, acting as phosphodiesterases, are involved in turnover of this secondary messenger and play a determinative role in the expression level of multicellular behavior in *S.* Typhimurium (Simm et al., 2007).

Thus, compared to planktonic growth, gene expression profile is different in biofilm cells and, most importantly, it not only depends on the temporal stage of biofilm development, but also on the spatial localization of the bacteria within the biofilm (Stewart and Franklin, 2008). Like in other bacteria, biofilm growth of *S.* Typhimurium has been shown to up-regulate genes involved in global regulation and tolerance to stress (Hamilton et al., 2009). Similarly, and under surface-associated growth, *S.* Enteritidis over-produces proteins mainly related to stress management (Giaouris et al., 2013). In another study, He and Ahn (2011) demonstrated, following culture of planktonic and biofilm cells of *S.* Typhimurium in tryptone soya broth (TSB), adjusted to pH 5.5 and pH 7.3, and gene expression evaluation, that the biofilm cells were more resistant to acidic stress compared with the planktonic cells; and the relative expression of toxin-, virulence-, efflux pump-related genes in the biofilm cells was distinct from that in the planktonic cells. Noteworthily, a number of *S. enterica* proteins have been recognized to be simultaneously implicated in stress response events, regulation of virulence, attachment processes and biofilm formation (deLivron and Robinson, 2008; Wang et al., 2010; Jennings et al., 2011; Salazar et al., 2013). Typical example is the alternative sigma factor sigma(S) (RpoS), which is required for virulence, stress resistance, biofilm formation, and development of the rdar morphotype (Ibanez-Ruiz et al., 2000; Robbe-Saule et al., 2006). Additionally, several diverse proteins can bind to the *csgD* promoter, including RpoS and the response regulator OmpR (Gerstel and Römling, 2003). CsgD is also regulated by the second messenger c-di-GMP at both transcriptional and post-transcriptional levels (Kader et al., 2006), while small non-coding RNAs have recently been shown to also be involved in its regulation (Van Puyvelde et al., 2013; Bordeau and Felden, 2014). From all the previous, it is obvious that the regulation of the rdar morphotype integrates a variety of complex regulatory systems which, obviously, should act in well organized hierarchical cascades.

A number of previous studies have revealed that the ability of *Salmonella* to attach to surfaces and create biofilms on them represent survival strategies leading to increased resistance to heat, acid, low relative humidity (RH), antibiotics and biocides (Joseph et al., 2001; Papavasileiou et al., 2010; Wan Norhana et al., 2010; Margas et al., 2014). *S.* Typhimurium cells presenting rugose phenotype and surrounded by exopolymeric substances, showed resistance to low pH and hydrogen peroxide and an ability to form biofilms (Anriany et al., 2001). In another study, *S.* Typhimurium embedded in a biofilm (pellicle) formed at the air-liquid interface resisted sodium hypochlorite at concentrations above 500 mg/L, while planktonic cells were sensitive to less than 50 mg/L (Scher et al., 2005). Interestingly, Amarasinghe et al. (2013) demonstrated that prolonged exposure of *S.* Typhimurium to a monoclonal polymeric IgA antibody (directed against the O antigen component of lipopolysaccharide), resulted in measurable exopolysaccharide accumulation and biofilm formation on both borosilicate glass surfaces and polystyrene microtiter plates.

Undoubtedly, from all the above it is clear that while building biofilms, cells physiologically adapt to sustain otherwise lethal impacts of a variety of environmental stress conditions. The objective of the next sections of this chapter is to provide an overview of the current knowledge related to the attachment and/or biofilm formation by *Salmonella* on both food contact and product surfaces in regard of the strong influence of these physical and inevitable phenomena on the enhancement of its survival within food processing.

2. Attachment and Biofilm Formation by *Salmonella* on Food Contact Surfaces and Survival

Bacterial attachment to either food contact surfaces or other abiotic materials is a complicated process, usually involving more than one mechanism and influenced by many factors, including the bacterial surface properties (physicochemical characteristics, surface biological structures), the material surface characteristics and the environmental factors, such as temperature, pH and culture conditions (e.g. presence and level of nutrients and/or of preconditioning films, static or flow conditions) (Palmer et al., 2007; Goulter et al., 2009; Van Houdt and Michiels, 2010). For instance, Nesse et al. (2014) found that hydrophobicity influenced the attachment of potentially human-pathogenic *Escherichia coli* from the ovine reservoir to different surfaces. Noteworthily, while strong bacterial attachment constitutes the first step of biofilm formation, attached bacterial cells do not necessarily proceed into sessile development. Thus, from the initial interaction with a substratum to the subsequent sessile growth, significant changes in expression of many genes should occur in the bacterial cells.

The molecular mechanisms by which bacteria are able to form biofilms in food processing plants are the subject of increasing interest in recent years and appear more complex than initially assumed (Hall-Stoodley et al., 2004; Smith et al., 2009). For instance, in an effort to characterize the molecular mechanism of attachment and biofilm formation of *S.* Typhimurium on food contact surfaces, Kim and Wei (2009) screened a library of random mutagenized clones for phenotypic analyses of their ability to form biofilm, pellicle, curli fimbriae and cellulose. Their findings revealed that many factors, such as production of extracellular polymeric substances (EPS) and their efficient transportation through outer membrane, expression of flagella and regulation of exo-ribonucleases and RNA binding proteins, could be involved in attachment and biofilm formation of this pathogen on food contact surfaces. Such knowledge may help develop better sanitizing procedures to be used in food processing.

2.1. Attachment by *Salmonella* to Food Contact Surfaces and Survival

Till now, numerous *in vitro* trials have shown that *Salmonella* can easily attach to various food-contact surfaces (such as stainless steel, plastic and glass) and survive for long periods (Giaouris et al., 2012), with the resistance afforded to these bacteria, once become attached to surfaces, to be documented in several of these studies. For instance, Finn et al. (2013b) studied the transcriptional responses of *S.* Typhimurium to desiccation on a stainless steel (SS) surface (at 24°C for 4 h, at *ca.* 45% RH). A total of 266 genes were differentially expressed under desiccation stress compared with a static broth culture, with the osmoprotectant transporters *proP*, *proU*, and *osmU* to be highly upregulated during drying. Deletion of any one of these transport systems resulted in a reduction in the long-term viability of *S.* Typhimurium on the SS surface. In another similar work studying the transcriptome of an auxotrophic *S.* Typhimurium mutant, desiccated on a plastic surface for 22 h, it was showed that the most highly transcribed genes were those involved in ribosomal structure, amino acid metabolism, energy production, ion transport, transcription, and stress response (Gruzdev et al., 2012). Bae et al. (2012) investigated the survival of *S.* Typhimurium in different forms (adhered cells, biofilm producing in TSB, biofilm producing at RH 100%) on the surface of SS and stored at various RHs, at room temperature for 5 days, and found that the survival varied depending on RH and attachment form. From these studies, it is clear that a diverse range of responses can be induced in response of attached cells to desiccation, depending on the precise experimental conditions involved (Finn et al., 2013a).

Noteworthily, it has been observed that acid acid-sensitive enteric pathogens, such as salmonellae, are protected from killing under extremely acidic conditions (e.g. of pH 2.5) when they are inoculated onto certain solid surfaces (Waterman and Small, 1998). Thus, inoculation onto fresh-cut produce surfaces, as well as onto inert surfaces, such as polyethersulfone membranes and tissue paper, significantly increased the survival of salmonellae during otherwise lethal acid challenge (Gawande and Bhagwat, 2002a,b). This surface-associated acid protection, which was observed in several *Salmonella* strains, required *de novo* protein synthesis and was independent of stationary-phase sigma transcription factor (encoded by *rpoS*).

Adsorption of macromolecules, such as food proteins, to solid surfaces leads to formation of conditioning films that may affect attachment of microorganisms and other cells. This adsorbed organic material can inhibit or facilitate attachment of arriving microorganisms and subsequently, once these succeed attachment, protect them from sanitizing procedures and, additionally, used as a potential source of nutrition. Biological soil (porcine serum) has been shown to increase the survival of *Salmonella* serovars attached to mortar surfaces, thus highlighting the need for effective cleaning to reduce harborage of these bacteria in the food factory environment, during otherwise lethal antimicrobial interventions (Allan et al., 2004). It's worth to be noted that starvation of *S.* Typhimurium cells, following their incubation in seawater for 3 years, has resulted in a significant decrease of the ability of starved cells to adhere to polystyrene, while on the other hand, this alarmingly increased the adhesion values to Hep2 cells (Lagha et al., 2012).

2.2. Biofilm Formation by *Salmonella* on Food Contact Surfaces and Survival

Biofilm matrix components, such as fimbriae and cellulose, have been suggested to play an important role in the survival of *Salmonella* in the environment (Solano et al., 2002). Based on these two components, multicellular behaviour by *Salmonella* is often categorized according to colony morphology into rdar (red, dry and rough) expressing curli fimbriae and cellulose, bdar (brown, dry and rough) expressing only curli fimbriae and pdar (pink, dry and rough) expressing only cellulose. Vestby et al. (2009a) investigated the distribution of morphotypes among 148 *Salmonella* strains of feed factory, clinical and reference collection origins, with emphasis on potential differences between morphotypes with regards to survival in the feed factory environment. Interestingly, chemical analysis showed no major differences in polysaccharide content between bdar and rdar strains, indicating that cellulose is not a major component of the *Salmonella* biofilm matrix. Additionally, the rdar and the bdar strains were found to be equally tolerant to disinfectants, with rdar morphology however appearing to be more favourable in long term survival in biofilm in a very dry environment. In some agreement, Chia et al. (2011) concluded that the rdar or bdar morphotype may influence persistence of poultry-associated *Salmonella* serovars, given that cellulose synthase gene null mutants displayed reduced attachment to abiotic surfaces relative to their parent strains.

Salmonella and some other Enterobacteriaceae possess the remarkable ability to survive in low-moisture environments for extended periods of time. Likely and as a results of this, many of the reported food-borne outbreaks associated with low-moisture foods involve *Salmonella* contamination. The mechanisms utilized by these bacteria to ensure their survival in dry conditions remain to be fully elucidated, however, in depth transcriptomic in combination with biofilm forming data is now beginning to emerge regarding this observation (Finn et al., 2013a). To this direction, Iibuchi et al. (2010) studied the survival of 17 *Salmonella* strains on polypropylene discs placed in a desiccation chamber and found that the biofilm-forming strains survived longer on the plastic surface than did biofilm-deficient strains. Authors suggested that the EPS on the surface may have protected the bacterial cells from dryness, although the possibility of some inherent resistance to environmental stresses linked to biofilm formation could not be excluded. Finn et al. (2013c) analyzed the phenotypes related to low-moisture survival of 46 *Salmonella* isolates and found that most of the isolates could form biofilms under defined laboratory conditions, with 57% being positive for curli fimbriae production and 75% being positive for cellulose production. Interestingly, food industrial biocides were found to be most effective against planktonic cells but less so when the same bacteria were surface dried or present as a biofilm. Cellulose-producing isolates were better survivors when exposed to a biocide compared with cellulose-negative isolates.

In order to figure out whether there is a correlation between biofilm forming ability of *Salmonella* and environmental persistence, Vestby et al. (2009b) investigated the abilities of 111 strains to form biofilm on polystyrene and in the air-liquid interface of nutrient broth. Significant differences were found between serovars, regarding the abilities, while when the strains were classified according to persistence, persistent strains produced more biofilm than non-persistent. These results indicate a correlation between persistence and biofilm formation suggesting that biofilm forming ability may be an important factor for persistence of *Salmonella* in the factory environment. In agreement with this assumption, a relation between biofilm forming capacity of 120 *Salmonella* strains isolated from multiple pork processing plants and survival on SS surfaces has also been revealed (Castelijn et al., 2013). Additionally, on such surfaces, biofilms showed greater and longer survival than planktonic cells, and they were also less susceptible to peracetic acid disinfection treatments. However, the latter effect was marginal and only observed in the presence of organic material (Castelijn et al., 2013). Rather contrary to these results, in another study with aim to investigate whether continuous contamination of light pasteurized egg products with *S.* Tennessee was caused by persistent contamination of the production facility, persistent isolates did not show a higher ability to form biofilm on a plastic surface than control strains (Jakočiūnė et al., 2014). In this study, the identified reason accounting for the persistent contamination was that the persistent strain had just adapted to grow in the relevant egg product. Schonewille et al. (2012) investigated the biofilm building capacity of different serotypes of poultry-derived *S. enterica* and concluded that this capacity is a function of adaptation to their host environment. Thus, the control of biofilms as a reservoir for *Salmonella* in the farm environment is of crucial importance for the overall improvement of food safety.

The effect of physiological state (planktonic *versus* biofilm) and prior exposure to desiccation and storage in dry milk powder on *Salmonella* survival and gene expression after passage through an *in vitro* digestion model was studied by Aviles et al. (2013). It was found that both the physiological state and length of storage affected the survival and gene expression of *Salmonella* within the desiccated milk powder environment and after passage through the *in vitro* digestion system, with the biofilms cells to present greater recovery than the planktonic ones. Additionally, increased expression of virulence-associated genes was seen in biofilm cells stored in milk powder, suggesting increased virulence potential. In another study on *Salmonella* biofilm formation, survival and virulence induction, Xu et al. (2010) characterized the viability and potential virulence of biofilm-forming *S.* Typhimurium under different pH levels, ranging from 5 to 7. At the lower pH values, the numbers of viable cells estimated by using real-time reverse transcription-PCR (RT-PCR) were greater than the culturable counts enumerated by a plate count method. Compared to planktonic cells, the *stn* and *invA* genes (coding for enterotoxin production and invasive ability, respectively) of biofilm cells were highly overexpressed in biofilm cells grown in pH 6. This result suggests that a regulator of virulence gene expression in biofilm cells may be triggered by different environmental conditions. This is in good agreement with a previous report showing the biofilm cells of *S.* Typhimurium to present higher adhesive ability in the spleens of mice (Turnock et al., 2002). Rodrigues et al. (2011) evaluated the susceptibility of *S. enterica* biofilms to four common food industry disinfectants, and analyzed how those chemical agents influenced stress-response and virulence genes expression by surviving cells. It was found that disinfection surviving biofilm cells seem to develop a stress response and/or become more virulent, which may compromise food safety and potentiate public health risk. Hopefully, all these results could provide a better understanding of both the survival and the pathogenicity of planktonic and biofilm *S.* Typhimurium cells under various stress conditions and valuable information for microbiological risk assessment.

Within a biofilm population, cells with diverse genotypes and phenotypes expressing distinct metabolic pathways, stress responses and other specific biological activities are juxtaposed (Stewart and Franklin, 2008). Indeed, a quite high number of studies dealt with and confirmed the enhancement of the survival of *Salmonella* biofilm cells exposed to various disinfectants compared to the same cells being planktonic. Tabak et al. (2007) investigated the susceptibility of planktonic and biofilm-associated *S.* Typhimurium to triclosan, a potent biocide that is included in a diverse range of products, with aim to identify potential mechanisms of adaptation. Biofilm-associated cells were quite more tolerant than planktonic cells (both log and stationary phases). However, bacteria derived from disrupted biofilms showed sensitivity to triclosan similar to stationary-phase planktonic cells. Based on supplementary gene expression results, authors suggested that *Salmonella* within biofilms could experience reduced influx, increased efflux and enhanced exopolysaccharides production. Er et al. (2014) determined the minimum inhibitory concentrations (MIC) and minimum biofilm inhibitory concentrations (MBIC) of selected preservatives against planktonic and biofilm forms of *Salmonella* and found that MBIC values were greater than the MIC values. Similar results have also been derived by Güven and Kaynak Onurdağ (2014), who determined the antimicrobial effects of some preservatives used in drug, cosmetics and food products and also compared the MBIC against biofilms of *S.* Typhimurium and some other microorganisms, formed on glass surfaces, and MIC values of the planktonic forms. O'Leary et al. (2013) characterized 172 *S.* Typhimurium isolates taken from the pork chain for their ability to form biofilms on SS and plastic surfaces and to survive chlorine-based challenges. It was found that the majority of strains readily attached to both type of surfaces, possessed biofilm forming capabilities and survived high chlorine concentrations. Somers et al. (1994) evaluated the effectiveness of trisodium phosphate (TSP), an approved food additive, against planktonic and biofilm cells of *S.* Typhimurium and other foodborne pathogens on SS and buna-N rubber materials, and found that biofilm cells of both *S.* Typhimurium and *L. monocytogenes* were more resistant than planktonic cells. Acidic and osmotic treatments are part of hurdle systems to control pathogens such as *Salmonella* in food. Previous results have shown that the ability of *Salmonella* strains to survive in the presence of acetic acid and rice vinegar parallels their ability to form biofilms (Hasegawa et al., 2011). The data obtained from all these studies indicate, if anything, that common preservatives and disinfectants used to control microbial growth in food environments and elsewhere, are not necessarily effective against biofilm microbial forms.

3. Attachment of *Salmonella* to Food Product Surfaces and Survival

Like in the case of bacterial attachment to food contact surfaces, attachment of bacteria to biotic food product surfaces depends on various factors, such as bacterial surface physicochemical properties and structures, type of food surfaces (e.g. meat, plant, tissue type and age, lean, fat) and environmental conditions (e.g. temperature, pH, presence of chemical substances, etc.). Numerous studies have attempted to explain this phenomenon as a function of bacterial cell surface and substratum physicochemical properties (i.e. hydrophobicity, charge), without however being always able to establish a strong relationship. It should not be neglected that bacterial surface contain proteins, polysaccharides, and other biopolymers that most likely affect its attachment to another surface (either biotic or abiotic), which is clearly a multifactorial process and further work is required to determine the full repertoire of the adhesive molecules involved (Jaglic et al., 2014).

3.1. Attachment of *Salmonella* to Produce Surfaces and Survival

Bacteria associated with plants have been observed to form assemblages referred to as aggregates, microcolonies, symplasmata, or biofilms on leaves, fruits, root surfaces and within intercellular spaces of plant tissues (Morris and Monier, 2003; Danhorn and Fuqua, 2007). Leaf internalization through stomata has been reported as a potential route of contamination, which renders food-borne pathogens protected from washing and disinfection by sanitizers (Golberg et al., 2011). *S. enterica* can rapidly adapt to environmental stresses and survive for long periods of time in various non-host habitats, including agricultural fields and the surface of fresh produce (Humphrey et al., 2004; Warriner and Namvar, 2010; Arthurson et al., 2011), leading, the last years, to major foodborne disease outbreaks (Hanning et al., 2009). Alarmingly, attachment and/or biofilm formation on plants surfaces have been shown to reduce the efficacy of sanitation treatments in food processing plants and also enhance stress resistance (Kyle et al., 2010; Niemira and Cooke, 2010). Kroupitski et al. (2009) investigated the interactions of *S. enterica* with abiotic and plant surfaces and their effect on the tolerance of the pathogen to various stressors. Strong biofilm producers were found to attach better to intact Romaine lettuce leaf tissue compared to weak producers. Exposure of lettuce-associated *Salmonella* cells to acidic conditions (pH 3.0) revealed increased tolerance of the attached *vs* planktonic bacteria. Several studies have focused on the survival, possible growth and/or internalization of *Salmonella* once inoculated onto the surface of fruits and vegetables and subsequently subjected to various treatments (Lang et al., 2004; Duffy et al., 2005; Daş et al., 2006; Iturriaga et al., 2003, 2007; Shi et al., 2007; Beuchat and Mann, 2008; Patel et al., 2013; Zheng et al., 2013). These studies have revealed the important influences of storage temperature, RH, inoculum size, source and inoculation method, serovar, biofilm forming ability, cultivar (plant genotype) and physiological state (including growth stage, ripeness and surface damage) of the plant, and interactions with resident bacteria in the phyllosphere, on bacteria viability and persistence. Strains associated with plants in particular have been shown to harbor specific traits that facilitate their adaptation to non-host environments. Such traits include the effective utilization of carbon sources abundant in plant tissues, the production of extracellular matrix, and biofilm formation (Fatica and Schneider, 2011; Martínez-Vaz et al., 2014).

The molecular mechanisms underlying the association of enteric pathogens with fresh produce are just beginning to be elucidated (Berger et al., 2010; Martínez-Vaz et al., 2014). Functional genomics studies have linked changes in gene expression to the bacterial attachment and colonization of plant structures. Thus, this research has shown that foodborne pathogens exhibit distinct changes in gene expression upon association with the surface of fresh produce. Interestingly, most genes involved in the molecular interactions of bacterial pathogens with produce are also implicated in cell surface structures, virulence, motility, and biofilm formation (Martínez-Vaz et al., 2014). Thus, the initial attachment to the plant is facilitated by cell surface structures important for surface attachment and biofilm formation. The subsequent survival in the harsh plant environment is mediated by the induction of stress responses, virulence factors and downregulation of genes involved in protein synthesis, energy, and metabolism (Figure 1).

To this direction, Brankatschk et al. (2013) investigated the transcriptional profile of *S.* Weltevreden during alfalfa sprout colonization and compared this with the corresponding when cells left to multiply planktonically in M9-glucose medium. Genes that were higher transcribed with sprouts were coding for proteins involved in mechanisms known to be important for attachment, motility and biofilm formation. Additionally, genes encoding structural and effector proteins of *Salmonella* pathogenicity island 2, involved in survival within macrophages during infection of animal tissue, were higher transcribed with sprouts, possibly as a response to environmental conditions.

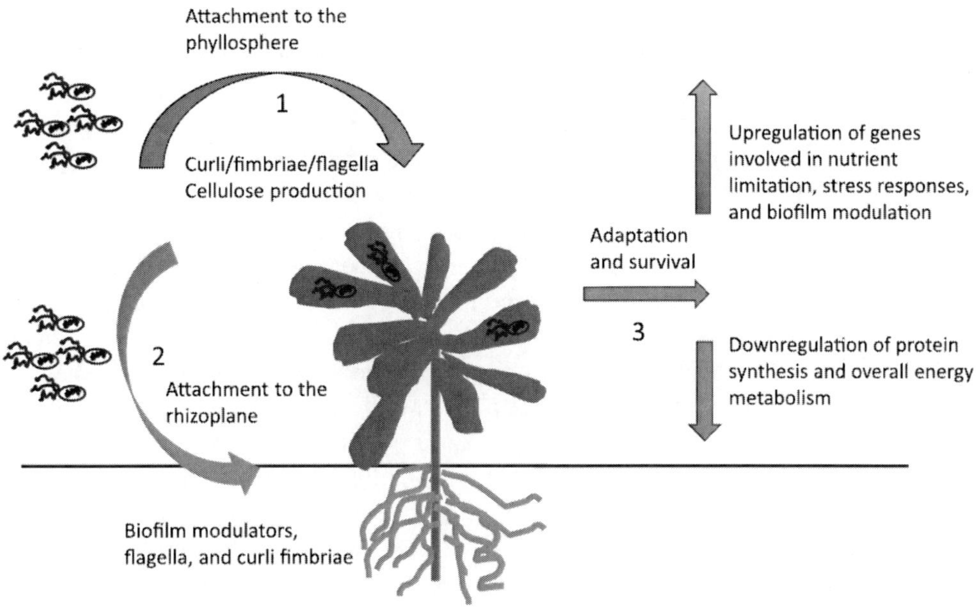

Figure 1. Molecular mechanisms mediating the colonization of leafy greens by human pathogens (figure obtained after permission from Martínez-Vaz et al., 2014, Enteric pathogen-plant interactions: molecular connections leading to colonization and growth and implications for food safety. *Microbes Environ.* 29, 123-135, Copyright © 1999- Japan Science and Technology Agency [JST]).

Nowadays, therefore it is increasingly becoming clear that the ability to produce biofilms together with the differential expression of genes involved in biofilm modulation, are characteristics linked to the association of certain human pathogens on produce. *S. enterica* is known to utilize diverse and overlapping strategies to interact with plants and their microflora, and to successfully colonize its vertebrate hosts (Brandl et al., 2013). Importantly, research has revealed that *S. enterica* genes important for virulence in animal systems (Fàbrega and Vila, 2013) are also required for colonization of plants (Barak et al., 2005; Kroupitski et al., 2013). Association of *Salmonella* with fresh produce appears to be serovar-specific involving flagella, curli, cellulose, and O antigen capsule (Barak et al., 2007; Lapidot and Yaron, 2009). These structures promote bacterial attachment to animal cells and play a critical role in biofilm formation and host colonization. Therefore, cellular structures used by *Salmonella* to invade animal hosts (Althouse et al., 2003; Kline et al., 2009) may also be important for their interaction and establishment in vegetables tissues. However, in another study, flagella and chemotaxis were found to be of minor importance for attachment of serovars Typhimurium and Dublin to plant leaves (Olsen et al., 2012). Understanding the interactions between *Salmonella* and plant surfaces might be useful in developing new approaches to prevent fresh produce-associated outbreaks.

Current industrial practice to decontaminate fresh produce involves the use of wash water supplemented with low concentrations of chlorine. Chlorine is an oxidative agent that is used to reduce the overall bacterial load on produce surface and prevent cross contamination in the wash water during minimal processing. However, studies have shown that *S. enterica* is capable of attaching to the produce surface and forming protective layers of biofilms (Brandl, 2006), making it difficult to completely inactivate the attached pathogens by chlorinated water washes alone. Lapidot et al. (2006) compared the adhesion and persistence on parsley of *S.* Typhimurium and its biofilm-deficient isogenic mutant (unable to produce cellulose and curli fimbriae). After a week of storage the biofilm producing strain survived chlorination significantly better than the biofilm-deficient mutant. However, the recovery of the mutant was still elevated, indicating that although the biofilm matrix has a role in persistence of *Salmonella* on leafy vegetables, this is not the most important mechanism, and other mechanisms could afford more significant protection effect. Salazar et al. (2013) investigated the functional roles of *ycfR*, *sirA*, and *yigG* in the attachment and survival of *S. enterica* on fresh produce (fresh spinach leaves and grape tomatoes) during procedures related to post-harvest minimal processing. Deletion of *ycfR* significantly reduced bacterial chlorine resistance and the attachment to the plant surfaces after chlorinated water washes. *ycfR* encodes a putative outer membrane protein in *E. coli* K-12 which regulates biofilm production via a process that involves changing surface properties of the bacterial cell and is also induced by several stress conditions (Zhang et al., 2007). In addition, they showed that both *sirA* which encodes a Type VI secretion protein, and *yigG*, which encodes a putative inner membrane protein, were also involved in *S. enterica* surface attachment and survival during post-harvest minimal processing of fresh produce.

Solutions containing calcium hypochlorite or sodium hypochlorite at concentrations of 1800 and 2000 micrograms/ml active (available) chlorine respectively, as well as 6% hydrogen peroxide or 80% ethanol were effective in reducing *Salmonella* populations on alfalfa seeds by more than 1000 fold (Beuchat, 1997). However, viable *Salmonella* cells were detected in seeds treated for 10 min in these solutions. The inaccessibility of *Salmonella* cells in crevices and between the cotyledon and testa of seeds to lethal concentrations of these chemicals are thought to be the reason for the lack of effectiveness. Jaquette et al. (1996) studied the efficacy of chlorine in killing *S.* Stanley inoculated onto alfalfa seeds and showed that chlorine concentrations of up to 1,040 mg/ml are not effective in eliminating the pathogen, although significant reduction in numbers of viable cells can be achieved. These results indicate that while populations of this pathogen can be greatly reduced by applying chlorine disinfection, elimination of this organism from alfalfa seeds may not be reliably achieved. This is even more worrying given the capability of survived cells to dramatically increase during the subsequent sprouting process and also retain their viability during refrigerated storage. On the contrary, although *S.* Typhimurium attached immediately to stone fruit surfaces when this was spot inoculated onto the sterile intact fructoplane of whole peaches and plums, attached cells were unable to survive simulated export conditions (0.5 ^{o}C) (Collignon and Korsten, 2010).

3.2. Attachment of *Salmonella* to Meat Surfaces and Survival

Meat and meat products have been identified as the most important source of transmission of *Salmonella* to humans. Microbial contamination of carcass surfaces occurs during slaughter and post slaughter processing steps, therefore interventions are needed to enhance meat safety and quality (Sofos et al., 1999; Giaouris et al., 2014). The animal tissues are surrounded by an extracellular matrix (ECM), which primarily consists of fibrous proteins (i.e. collagen, fibronectin, laminin, and elastin) and proteoglycans. Due to its ubiquitous nature, the ECM also serves as a substrate for microbial attachment (Selgas et al., 1993). Although the attachment to meat surfaces has been shown to protect *Salmonella* cells from various stresses, little is known about the specific processes that influence this process. Morild et al. (2011) studied the attachment of *S.* Typhimurium, *L. monocytogenes* and *Yersinia enterocolitica* to pig skin and muscle tissue. Attachment properties differed between skin and muscle surfaces, while no significant differences in surface attachment between the three bacterial species were observed. Microscopy studies showed that bacteria were located in deep tissue structures of skin and muscle surfaces. Interestingly, a significantly higher number of firmly attached bacteria were found on the skin surface following decontamination with hot water or lactic acid, compared to the untreated skin surface. The attachment of *S.* Typhimurium to pork muscle tissue was found to increase its heat tolerance (Humphrey et al., 1997), while the attachment to poultry skin increased its ability to resist various disinfectants and this efficacy was influenced by extent of attachment of bacteria to skin and method of disinfectant application (Tamblyn et al., 1997). Some other studies have evaluated the viability of *S.* Typhimurium inoculated onto the surface of some ready-to-eat meat products, without, however, concluding something alarming (Porto-Fett et al., 2008a,b; Jacob et al., 2009).

S. Typhimurium DT104 strain has emerged as a global human and veterinary public health concern because of its antibiotic resistance and extensive host range. Kinsella et al. (2007) investigated the influence of attachment to beef surfaces on the survival, injury and death of stationary phase cells of *S.* Typhimurium DT104, at various a_w values and temperatures, compared to cells free in solution. They found that attachment of cells to meat surfaces prevented cell injury and death from hyperosmosis and low temperatures, compared to cells in meat juices. However, after storage of cells for 72 h, cell injury was reduced and viability increased in meat juices, compared to cells attached to meat surfaces. In another similar study aiming to investigate the survival of S*almonella* on meat surfaces, Kinsella et al. (2009) recorded changes in *S.* Typhimurium DT104 and total viable count (TVC) survival on meat pieces (taken from different sites on beef carcasses) stored for 72 h under different combinations of RH (i.e. 75% or 96%) and temperature (i.e. 5°C or 10°C). *Salmonella* counts declined during storage at low RH conditions at both temperatures, while these increased during storage at high RH at 10°C only.

Conclusion

From an ecological point of view, food processing environments could be considered microbial habitats that are constantly disturbed by sanitizing procedures. Obviously, the best strategy to improve food safety is the adequate hygiene together with the application of antimicrobial intervention technologies at all stages of the food chain. However, the complete elimination of unwanted bacteria from such environments is a rather difficult task, because these can attach to both food and processing equipment surfaces and form biofilms, where they persist even after cleaning and disinfection (Frank, 2001). Biofilm bacteria can cause a range of problems in food industry including reduced heat-cold transfer, clogging water pipelines, food spoilage and they may cause infections among consumers. Attached bacterial cells are more resilient to displacement strategies including physical and chemical procedures, as a result of their more resistant phenotype compared to planktonic counterparts.

The persistence of bacterial pathogens within food environments is of great importance; while many contributory factors may exist that collectively determine this ability, such as attachment strength, biofilm forming potential, antimicrobial resistance and survival ability. The deep understanding of the mechanisms facilitating bacterial attachment to surfaces is desired by numerous industries including those occupied with food production and processing. Future biofilm investigations with the aim of potential control approaches should include a combination of bacterial genetics, systems biology, mechanical engineering and biological chemistry (Tan et al., 2014).

Extra-animal survival is an important parameter for the environmental dissemination of zoonotic salmonellae, with the ability of these bacteria to survive in the food chain to be largely due to their skill to effectively respond to various environmental changes (Shen and Fang, 2012; Spector and Kenyon, 2012; Runkel et al., 2013). A quite high number of previous studies aiming to explain stress adaptation and survival mechanisms of *Salmonella* have linked these processes to the ability of bacteria to attach to surfaces and create biofilms. Part of these studies, related to attachment and biofilm formation on food contact and product surfaces, has been reviewed in this chapter. Results presented verify the recalcitrance of biofilm communities and highlighted a food safety subject which is not to be ignored in food processing. Attachment, colonization, as well as biofilm formation capacity of *Salmonella* on such surfaces may cause continuous contamination. It is now becoming clear that biofilms may play a crucial role in the survival of salmonellae under unfavorable environmental conditions, such as in animal slaughterhouses, farms and food processing plants, and could serve as a reservoir compromising food safety. It is unlikely that *Salmonella* will ever be eradicated from the food chain, therefore the results from laboratory research, such as investigations on the response of *Salmonella* to different stresses under surface associated growth, should be translated into improved intervention strategies for food producers and consumers.

A greater understanding of the ability of *Salmonella* to survive in the environment is now necessary to effectively control this pathogen. Future studies should focus on the selection of the best strains to study survival mechanisms, investigation of different environmental conditions and practices and their influence on resistance and examination of *Salmonella*'s molecular responses to stresses to ascertain the simultaneous roles of specific genes on both surface attachment and stress resistance. The uncovering of the regulatory networks and metabolic pathways engaged in the course of sessile growth is much promising. An improved understanding of the physiological responses taking place inside a *Salmonella* biofilm, can be of value to work out the relative roles of benefits and forces that drive the switch to this sessile mode of growth. Much work is also needed to examine the mechanisms by which exposure to stress and/or biofilm formation contribute to virulence, especially interactions between *Salmonella* and the host. Bacterial biofilms generally are more resistant to stresses as compared with free planktonic cells. Therefore, the discovery of antimicrobial stress factors that have strong inhibitory effects on bacterial biofilm formation is wanted. Remarkably, in food processing environments, a variety of different bacteria may attach to surfaces, survive, grow and sometimes form multispecies sessile communities (Jahid and Ha, 2014). Survival and response of *Salmonella* in complex polymicrobial biofilms should also be examined as cell-to-cell interactions may influence gene expression and stress resistance (Moons et al., 2009; Wang et al., 2013). All this research is important for more informed assessment of microbial risk and hopefully will lead to the development of more effective and targeted strategies to minimize the contamination with this pathogen in food processing, its attachment to surfaces and biofilm formation on them, with a great positive impact on food safety and public health.

Acknowledgment

Authors E.G. and L.L.N are members of the EU COST Action FA1202: A European Network for Mitigating Bacterial Colonisation and Persistence on Foods and Food Processing Environments (http://www.bacfoodnet.org/) and acknowledge this action for facilitating collaborative networking that assisted in the writing of this chapter.

References

Ahmad, I., Lamprokostopoulou, A., Le Guyon, S., Streck, E., Barthel, M., Peters, V., Hardt, W. D., and Römling, U. (2011). Complex c-di-GMP signaling networks mediate transition between virulence properties and biofilm formation in *Salmonella enterica* serovar Typhimurium. *PLoS One* 6(12):e28351.

Allan, J. T., Yan, Z., Genzlinger, L. L., and Kornacki, J. L. (2004). Temperature and biological soil effects on the survival of selected foodborne pathogens on a mortar surface. *J. Food Prot.* 67, 2661-2665.

Althouse, C., Patterson, S., Fedorka-Cray, P., and Isaacson, R. E. (2003). Type 1 fimbriae of *Salmonella enterica* serovar Typhimurium bind to enterocytes and contribute to colonization of swine *in vivo*. *Infect. Immun.* 71, 6446-6452.

Amarasinghe, J. J., D'Hondt, R. E., Waters, C. M., and Mantis, N. J. (2013). Exposure of *Salmonella enterica* serovar Typhimurium to a protective monoclonal IgA triggers exopolysaccharide production via a diguanylate cyclase-dependent pathway. *Infect. Immun.* 81, 653-664.

Anderson, G. G., and O'Toole, G. A. (2008). Innate and induced resistance mechanisms of bacterial biofilms. *Curr. Top. Microbiol. Immunol.* 322, 85-105.

Anriany, Y. A., Weiner, R. M., Johnson, J. A., De Rezende, C. E., and Joseph, S. W. (2001). *Salmonella enterica* serovar Typhimurium DT104 displays a rugose phenotype. *Appl. Environ. Microbiol.* 67, 4048-4056.

Arthurson, V., Sessitsch, A., and Jäderlund, L. (2011). Persistence and spread of *Salmonella enterica* serovar Weltevreden in soil and on spinach plants. *FEMS Microbiol. Lett.* 314, 67-74.

Audia, J. P., Webb, C. C., and Foster, J. W. (2001). Breaking through the acid barrier: an orchestrated response to proton stress by enteric bacteria. *Int. J. Med. Microbiol.* 291, 97-106.

Aviles, B., Klotz, C., Eifert, J., Williams, R., and Ponder, M. (2013). Biofilms promote survival and virulence of *Salmonella enterica* sv. Tennessee during prolonged dry storage and after passage through an *in vitro* digestion system. *Int. J. Food Microbiol.* 162, 252-259.

Bae, Y. M., Baek, S. Y., and Lee, S. Y. (2012). Resistance of pathogenic bacteria on the surface of stainless steel depending on attachment form and efficacy of chemical sanitizers. *Int. J. Food Microbiol.* 153, 465-473.

Barak, J. D., Gorski, L., Naraghi-Arani, P., and Charkowski, A. O. (2005). *Salmonella enterica* virulence genes are required for bacterial attachment to plant tissue. *Appl. Environ. Microbiol.* 71, 5685-5691.

Barak, J. D., Jahn, C. E., Gibson, D. L., and Charkowski, A. O. (2007). The role of cellulose and O-antigen capsule in the colonization of plants by *Salmonella enterica*. *Mol. Plant Microbe Interact.* 20, 1083-1091.

Berger, C. N., Sodha, S. V., Shaw, R. K., Griffin, P. M., Pink, D., Hand, P., and Frankel, G. (2010). Fresh fruit and vegetables as vehicles for the transmission of human pathogens. *Environ. Microbiol.* 12, 2385-2397.

Beuchat, L. R. (1997). Comparison of chemical treatments to kill *Salmonella* on alfalfa seeds destined for sprout production. *Int. J. Food Microbiol.* 34, 329-333.

Beuchat, L. R., and Mann, D. A. (2008). Survival and growth of acid-adapted and unadapted *Salmonella* in and on raw tomatoes as affected by variety, stage of ripeness, and storage temperature. *J. Food Prot.* 71, 1572-1579.

Bordeau, V., and Felden, B. (2014). Curli synthesis and biofilm formation in enteric bacteria are controlled by a dynamic small RNA module made up of a pseudoknot assisted by an RNA chaperone. *Nucleic Acids Res.* 42, 4682-4696.

Brandl, M. T. (2006). Fitness of human enteric pathogens on plants and implications for food safety. *Annu. Rev. Phytopathol.* 44, 367-392.

Brandl, M. T., Cox, C. E., and Teplitski, M. (2013). *Salmonella* interactions with plants and their associated microbiota. *Phytopathology* 103, 316-325.

Brankatschk, K., Kamber, T., Pothier, J. F., Duffy, B., and Smits, T. H. (2013). Transcriptional profile of *Salmonella enterica* subsp. *enterica* serovar Weltevreden during alfalfa sprout colonization. *Microb. Biotechnol.* in press doi: 10.1111/1751-7915.12104.

Castelijn, G. A., Parabirsing, J. A., Zwietering, M. H., Moezelaar, R., and Abee, T. (2013). Surface behaviour of *S.* Typhimurium, *S.* Derby, *S.* Brandenburg and *S.* Infantis. *Vet. Microbiol.* 161, 305-314.

Chia, T. W., McMeekin, T. A., Fegan, N., and Dykes, G. A. (2011). Significance of the rdar and bdar morphotypes in the hydrophobicity and attachment to abiotic surfaces of *Salmonella* Sofia and other poultry-associated *Salmonella* serovars. *Lett. Appl. Microbiol.* 53, 581-584.

Chmielewski, R. A., and Frank, J. F. (2003). Biofilm formation and control in food processing facilities. *Compr. Rev. Food Sci. Food Saf.* 2, 22–32.

Coenye, T. (2010). Response of sessile cells to stress: from changes in gene expression to phenotypic adaptation. *FEMS Immunol. Med. Microbiol.* 59, 239-252.

Collignon, S., and Korsten, L. (2010). Attachment and colonization by *Escherichia coli* O157:H7, *Listeria monocytogenes*, *Salmonella enterica* subsp. enterica serovar Typhimurium, and *Staphylococcus aureus* on stone fruit surfaces and survival through a simulated commercial export chain. *J. Food Prot.* 73, 1247-1256.

Cotter, P. A., and Stibitz, S. (2007). c-di-GMP-mediated regulation of virulence and biofilm formation. *Curr. Opin. Microbiol.* 10, 17-23.

Danhorn, T., and Fuqua, C. (2007). Biofilm formation by plant-associated bacteria. *Annu. Rev. Microbiol.* 61, 401-422.

Daş, E., Gürakan, G. C., and Bayindirli, A. (2006). Effect of controlled atmosphere storage, modified atmosphere packaging and gaseous ozone treatment on the survival of *Salmonella* Enteritidis on cherry tomatoes. *Food Microbiol.* 23, 430-438.

deLivron, M. A., and Robinson, V. L. (2008). *Salmonella enterica* serovar Typhimurium BipA exhibits two distinct ribosome binding modes. *J. Bacteriol.* 190, 5944-5952.

Dorman, C. J. (2009). Global regulators and environmental adaptation in Gram-negative pathogens. *Clin. Microbiol. Infect.* 15 Suppl 1, 47-50.

Duffy, E. A., Cisneros-Zevallos, L., Castillo, A., Pillai, S. D., Ricke, S. C., and Acuff, G. R. (2005). Survival of *Salmonella* transformed to express green fluorescent protein on Italian parsley as affected by processing and storage. *J. Food Prot.* 68, 687-695.

Duguid, J. P., Anderson, E. S., and Campbell, I. (1966). Fimbriae and adhesive properties in salmonellae. *J. Pathol. Bacteriol.* 92, 107-138.

Er, B., Demirhan, B., Onurdag, F. K., Ozgacar, S. Ö., and Oktem, A. B. (2014). Antimicrobial and antibiofilm effects of selected food preservatives against *Salmonella* spp. isolated from chicken samples. *Poult. Sci.* 93, 695-701.

Fàbrega, A., and Vila, J. (2013). *Salmonella enterica* serovar Typhimurium skills to succeed in the host: virulence and regulation. *Clin. Microbiol. Rev.* 26, 308-341.

Fatica, M. K., and Schneider, K. R. (2011). *Salmonella* and produce: survival in the plant environment and implications in food safety. *Virulence* 2, 573-579.

Finn, S., Condell, O., McClure, P., Amézquita, A., and Fanning, S. (2013a). Mechanisms of survival, responses and sources of *Salmonella* in low-moisture environments. *Front Microbiol.* 4:331.

Finn, S., Händler, K., Condell, O., Colgan, A., Cooney, S., McClure, P., Amézquita, A., Hinton, J. C., and Fanning, S. (2013b). ProP is required for the survival of desiccated *Salmonella enterica* serovar Typhimurium cells on a stainless steel surface. *Appl. Environ. Microbiol.* 79, 4376-4384.

Finn, S., Hinton, J. C., McClure, P., Amézquita, A., Martins, M., and Fanning, S. (2013c). Phenotypic characterization of *Salmonella* isolated from food production environments associated with low-water activity foods. *J. Food Prot.* 76, 1488-1499.

Foley, S. L., Johnson, T. J., Ricke, S. C., Nayak, R., and Danzeisen, J. (2013). *Salmonella* pathogenicity and host adaptation in chicken-associated serovars. *Microbiol. Mol. Biol. Rev.* 77, 582-607.

Foster, J. W., and Spector, M. P. (1995). How *Salmonella* survive against the odds. *Annu. Rev. Microbiol.* 49, 145-174.

Frank, J. F. (2001). Microbial attachment to food and food contact surfaces. *Adv. Food Nutr. Res.* 43, 319-370.

Fux, C. A., Costerton, J. W., Stewart, P. S., and Stoodley, P. (2005). Survival strategies of infectious biofilms. *Trends Microbiol.* 13, 34-40.

Gawande, P. V., and Bhagwat, A. A. (2002a). Inoculation onto solid surfaces protects *Salmonella* spp. during acid challenge: a model study using polyethersulfone membranes. *Appl. Environ. Microbiol.* 68, 86-92.

Gawande, P. V., and Bhagwat, A. A. (2002b). Protective effects of cold temperature and surface-contact on acid tolerance of *Salmonella* spp.. *J. Appl. Microbiol.* 93, 689-696.

Gerstel, U., Park, C., and Römling, U. (2003). Complex regulation of *csgD* promoter activity by global regulatory proteins. *Mol. Microbiol.* 49, 639-654.

Gerstel, U., and Römling, U. (2001). Oxygen tension and nutrient starvation are major signals that regulate *agfD* promoter activity and expression of the multicellular morphotype in *Salmonella* Typhimurium. *Environ. Microbiol.* 3, 638-648.

Gerstel, U., and Römling, U. (2003). The *csgD* promoter, a control unit for biofilm formation in *Salmonella typhimurium*. *Res. Microbiol.* 154, 659-667.

Giaouris, E., Chorianopoulos, N., Skandamis, P. and Nychas, G.-J. (2012). Attachment and biofilm formation by *Salmonella* in food processing environments. *In* B.S.M. Mahmoud (Ed.), *Salmonella*: A dangerous foodborne pathogen (pp. 157–180), Intech Open Access Publisher.

Giaouris, E., Heir, E., Hébraud, M., Chorianopoulos, N., Langsrud, S., Møretrø, T., Habimana, O., Desvaux, M., Renier, S., and Nychas, G. J. (2014). Attachment and biofilm formation by foodborne bacteria in meat processing environments: causes, implications, role of bacterial interactions and control by alternative novel methods. *Meat Sci.* 97, 298-309.

Giaouris, E., Samoilis, G., Chorianopoulos, N., Ercolini, D., and Nychas, G. J. (2013). Differential protein expression patterns between planktonic and biofilm cells of *Salmonella enterica* serovar Enteritidis PT4 on stainless steel surface. *Int. J. Food Microbiol.* 162, 105-113.

Gilbert, P., Allison, D. G., and McBain, A. J. (2002a). Biofilms *in vitro* and *in vivo*: do singular mechanisms imply cross-resistance? *J. Appl. Microbiol.* 92, 98S-110S.

Gilbert, P., Maira-Litran, T., McBain, A. J., Rickard, A. H., and Whyte, F. W. (2002b). The physiology and collective recalcitrance of microbial biofilm communities. *Adv. Microb. Physiol.* 46, 202-256.

Golberg, D., Kroupitski, Y., Belausov, E., Pinto, R., and Sela, S. (2011). *Salmonella* Typhimurium internalization is variable in leafy vegetables and fresh herbs. *Int. J. Food Microbiol.* 145, 250-257.

Goulter, R. M., Gentle, I. R., and Dykes, G. A. (2009). Issues in determining factors influencing bacterial attachment: a review using the attachment of *Escherichia coli* to abiotic surfaces as an example. *Lett. Appl. Microbiol.* 49, 1-7.

Gounadaki, A. S., Skandamis, P. N., Drosinos, E. H., and Nychas, G. J. (2008). Microbial ecology of food contact surfaces and products of small-scale facilities producing traditional sausages. *Food Microbiol.* 25, 313-323.

Gruzdev, N., McClelland, M., Porwollik, S., Ofaim, S., Pinto, R., and Saldinger-Sela, S. (2012). Global transcriptional analysis of dehydrated *Salmonella enterica* serovar Typhimurium. *Appl. Environ. Microbiol.* 78, 7866-7875.

Güven, N., and Kaynak Onurdağ, F. (2014). Investigation of antimicrobial and antibiofilm effects of some preservatives used in drugs, cosmetics and food products. *Mikrobiyol. Bul.* 48, 94-105.

Hall-Stoodley, L., Costerton, J. W., and Stoodley, P. (2004). Bacterial biofilms: from the natural environment to infectious diseases. *Nat. Rev. Microbiol.* 2, 95-108.

Hamilton, S., Bongaerts, R. J., Mulholland, F., Cochrane, B., Porter, J., Lucchini, S., Lappin-Scott, H. M., and Hinton, J. C. (2009). The transcriptional programme of *Salmonella enterica* serovar Typhimurium reveals a key role for tryptophan metabolism in biofilms. *BMC Genomics* 10:599.

Hammar, M., Arnqvist, A., Bian, Z., Olsén, A., and Normark, S. (1995). Expression of two *csg* operons is required for production of fibronectin- and congo red-binding curli polymers in *Escherichia coli* K-12. *Mol. Microbiol.* 18, 661-670.

Hanning, I. B., Nutt, J. D., and Ricke, S. C. (2009). Salmonellosis outbreaks in the United States due to fresh produce: sources and potential intervention measures. *Foodborne Pathog. Dis.* 6, 635-648.

Hasegawa, A., Hara-Kudo, Y., and Kumagai, S. (2011). Survival of *Salmonella* strains differing in their biofilm-formation capability upon exposure to hydrochloric and acetic acid and to high salt. *J. Vet. Med. Sci.* 73, 1163-1168.

He, X., and Ahn, J. (2011). Differential gene expression in planktonic and biofilm cells of multiple antibiotic-resistant *Salmonella* Typhimurium and *Staphylococcus aureus*. *FEMS Microbiol. Lett.* 325, 180-188.

Humphrey, T. (2004). *Salmonella*, stress responses and food safety. *Nat. Rev. Microbiol.* 2, 504-509.

Humphrey, T. J., Wilde, S. J., and Rowbury, R. J. (1997). Heat tolerance of *Salmonella* Typhimurium DT104 isolates attached to muscle tissue. *Lett. Appl. Microbiol.* 25, 265-268.

Ibanez-Ruiz, M., Robbe-Saule, V., Hermant, D., Labrude, S., and Norel, F. (2000). Identification of RpoS (sigma(S))-regulated genes in *Salmonella enterica* serovar Typhimurium. *J. Bacteriol.* 20, 5749-5756.

Iibuchi, R., Hara-Kudo, Y., Hasegawa, A., and Kumagai, S. (2010). Survival of *Salmonella* on a polypropylene surface under dry conditions in relation to biofilm-formation capability. *J. Food Prot.* 73, 1506-1510.

Iturriaga, M. H., Escartín, E. F., Beuchat, L. R., and Martínez-Peniche, R. (2003). Effect of inoculum size, relative humidity, storage temperature, and ripening stage on the attachment of *Salmonella* Montevideo to tomatoes and tomatillos. *J. Food Prot.* 66, 1756-1761.

Iturriaga, M. H., Tamplin, M. L., and Escartín, E. F. (2007). Colonization of tomatoes by *Salmonella* Montevideo is affected by relative humidity and storage temperature. *J. Food Prot.* 70, 30-34.

Jacob, R., Porto-Fett, A. C., Call, J. E., and Luchansky, J. B. (2009). Fate of surface-inoculated *Escherichia coli* O157:H7, *Listeria monocytogenes*, and *Salmonella* Typhimurium on kippered beef during extended storage at refrigeration and abusive temperatures. *J. Food Prot.* 72, 403-407.

Jaglic, Z., Desvaux, M., Weiss, A., Nesse, L. L., Meyer, R. L., Demnerova, K., Schmidt, H., Giaouris, E., Sipailiene, A., Teixeira, P., Kačániová, M., Riedel, C., and Knøchel, S.

(2014). Surface adhesins and exopolymers of selected foodborne pathogens. *Microbiology* in press doi: 10.1099/mic.0.075887-0.

Jahid, I. K., and Ha, S.-D. (2014). The Paradox of Mixed-Species Biofilms in the Context of Food Safety. *Compr. Rev. Food. Sci. Food Saf.* 13, 990-1011.

Jakočiūnė, D., Bisgaard, M., Pedersen, K., and Olsen, J. E. (2014). Demonstration of persistent contamination of a cooked egg product production facility with *Salmonella enterica* serovar Tennessee and characterization of the persistent strain. *J. Appl. Microbiol.* 117, 547-553.

Jaquette, C. B., Beuchat, L. R., and Mahon, B. E. (1996). Efficacy of chlorine and heat treatment in killing *Salmonella* Stanley inoculated onto alfalfa seeds and growth and survival of the pathogen during sprouting and storage. *Appl. Environ. Microbiol.* 62, 2212-2215.

Jennings, M. E., Quick, L. N., Soni, A., Davis, R. R., Crosby, K., Ott, C. M., Nickerson, C. A., and Wilson, J. W. (2011). Characterization of the *Salmonella enterica* serovar Typhimurium *ydcI* gene, which encodes a conserved DNA binding protein required for full acid stress resistance. *J. Bacteriol.* 193, 2208-2217.

Joseph, B., Otta, S. K., Karunasagar, I., and Karunasagar, I. (2001). Biofilm formation by *Salmonella* spp. on food contact surfaces and their sensitivity to sanitizers. *Int. J. Food Microbiol.* 64, 367-372.

Kader, A., Simm, R., Gerstel, U., Morr, M., and Römling, U. (2006). Hierarchical involvement of various GGDEF domain proteins in rdar morphotype development of *Salmonella enterica* serovar Typhimurium. *Mol. Microbiol.* 60, 602-616.

Kim, S. H., and Wei, C. I. (2009). Molecular characterization of biofilm formation and attachment of *Salmonella enterica* serovar Typhimurium DT104 on food contact surfaces. *J. Food Prot.* 72, 1841-1847.

Kinsella, K. J., Prendergast, D. M., McCann, M. S., Blair, I. S., McDowell, D. A., and Sheridan, J. J. (2009). The survival of *Salmonella enterica* serovar Typhimurium DT104 and total viable counts on beef surfaces at different relative humidities and temperatures. *J. Appl. Microbiol.* 106, 171-180.

Kinsella, K. J., Rowe, T. A., Blair, I. S., McDowell, D. A., and Sheridan, J. J. (2007). The influence of attachment to beef surfaces on the survival of cells of *Salmonella enterica* serovar Typhimurium DT104, at different a(w) values and at low storage temperatures. *Food Microbiol.* 24, 786-793.

Kline, K. A., Fälker, S., Dahlberg, S., Normark, S., and Henriques-Normark, B. (2009). Bacterial adhesins in host-microbe interactions. *Cell Host Microbe* 5, 580-592.

Kroupitski, Y., Brandl, M. T., Pinto, R., Belausov, E., Tamir-Ariel, D., Burdman, S., and Sela Saldinger, S. (2013). Identification of *Salmonella enterica* genes with a role in persistence on lettuce leaves during cold storage by recombinase-based *in vivo* expression technology. *Phytopathology* 103, 362-372.

Kroupitski, Y., Pinto, R., Brandl, M. T., Belausov, E., and Sela, S. (2009). Interactions of *Salmonella enterica* with lettuce leaves. *J. Appl. Microbiol.* 106, 1876-1885.

Kyle, J. L., Parker, C. T., Goudeau, D., and Brandl, M. T. (2010). Transcriptome analysis of *Escherichia coli* O157:H7 exposed to lysates of lettuce leaves. *Appl. Environ. Microbiol.* 76, 1375-1387.

Lagha, R., Ben Abdallah, F., Ellafi, A., Békir, K., and Bakhrouf, A. (2012). Biofilm formation, cell surface hydrophobicity, and fatty acids analysis of starved *Salmonella enterica* serovar Typhimurium in seawater. *Foodborne Pathog. Dis.* 9, 786-791.

Lang, M. M., Harris, L. J., and Beuchat, L. R. (2004). Survival and recovery of *Escherichia coli* O157:H7, *Salmonella*, and *Listeria monocytogenes* on lettuce and parsley as affected by method of inoculation, time between inoculation and analysis, and treatment with chlorinated water. *J. Food Prot.* 67, 1092-1103.

Lapidot, A., Romling, U., and Yaron, S. (2006). Biofilm formation and the survival of *Salmonella* Typhimurium on parsley. *Int. J. Food Microbiol.* 109, 229-233.

Lapidot, A., and Yaron, S. (2009). Transfer of *Salmonella enterica* serovar Typhimurium from contaminated irrigation water to parsley is dependent on curli and cellulose, the biofilm matrix components. *J. Food Prot.* 72, 618-623.

Le Guyon, S., Simm, R., Rhen, M., and Römling, U. (2014). Dissecting the c-di-GMP signaling network regulating motility in *Salmonella enterica* serovar Typhimurium. *Environ. Microbiol.* in press, doi: 10.1111/1462-2920.12580.

Mah, T. F., and O'Toole, G. A. (2001). Mechanisms of biofilm resistance to antimicrobial agents. *Trends Microbiol.* 9, 34-39.

Mangalappalli-Illathu, A. K., Lawrence, J. R., and Korber, D. R. (2009). Cells in shearable and nonshearable regions of *Salmonella enterica* serovar Enteritidis biofilms are morphologically and physiologically distinct. *Can. J. Microbiol.* 55, 955-966.

Margas, E., Meneses, N., Conde-Petit, B., Dodd, C. E., and Holah, J. (2014). Survival and death kinetics of *Salmonella* strains at low relative humidity, attached to stainless steel surfaces. *Int. J. Food Microbiol.* 187, 33-40.

Martínez-Vaz, B. M., Fink, R. C., Diez-Gonzalez, F., and Sadowsky, M. J. (2014). Enteric pathogen-plant interactions: molecular connections leading to colonization and growth and implications for food safety. *Microbes Environ.* 29, 123-135.

Moons, P., Michiels, C. W., and Aertsen, A. (2009). Bacterial interactions in biofilms. *Crit. Rev. Microbiol.* 35, 157-168.

Morild, R. K., Olsen, J. E., and Aabo, S. (2011). Change in attachment of *Salmonella* Typhimurium, *Yersinia enterocolitica*, and *Listeria monocytogenes* to pork skin and muscle after hot water and lactic acid decontamination. *Int. J. Food Microbiol.* 145, 353-358.

Morris, C. E., and Monier, J. M. (2003). The ecological significance of biofilm formation by plant-associated bacteria. *Annu. Rev. Phytopathol.* 41, 429-453.

Nesse, L. L., Nordby, K., Heir, E., Bergsjoe, B., Vardund, T., Nygaard, H., and Holstad, G. (2003). Molecular analyses of *Salmonella enterica* isolates from fish feed factories and fish feed ingredients. *Appl. Environ. Microbiol.* 69, 1075-1081.

Nesse, L. L., Sekse, C., Berg, K., Johannesen, K. C., Solheim, H., Vestby, L. K., and Urdahl, A. M. (2014). Potentially pathogenic *Escherichia coli* can form a biofilm under conditions relevant to the food production chain. *Appl. Environ. Microbiol.* 80, 2042-2049.

Niemira, B. A., and Cooke, P. H. (2010). *Escherichia coli* O157:H7 biofilm formation on Romaine lettuce and spinach leaf surfaces reduces efficacy of irradiation and sodium hypochlorite washes. *J. Food Sci.* 75, M270- M277.

O'Leary, D., Cabe, E. M., McCusker, M. P., Martins, M., Fanning, S., and Duffy, G. (2013). Microbiological study of biofilm formation in isolates of *Salmonella enterica*

Typhimurium DT104 and DT104b cultured from the modern pork chain. *Int. J. Food Microbiol.* 161, 36-43.

Olsen, J. E., Hoegh-Andersen, K. H., Casadesús, J., and Thomsen, L. E. (2012). The importance of motility and chemotaxis for extra-animal survival of *Salmonella enterica* serovar Typhimurium and Dublin. *J. Appl. Microbiol.* 113, 560-568.

Palmer, J., Flint, S., and Brooks, J. (2007). Bacterial cell attachment, the beginning of a biofilm. *J. Ind. Microbiol. Biotechnol.* 34, 577-588.

Papavasileiou, K., Papavasileiou, E., Tseleni-Kotsovili, A., Bersimis, S., Nicolaou, C., Ioannidis, A., and Chatzipanagiotou, S. (2010). Comparative antimicrobial susceptibility of biofilm versus planktonic forms of *Salmonella enterica* strains isolated from children with gastroenteritis. *Eur. J. Clin. Microbiol. Infect. Dis.* 29, 1401-1405.

Patel, J., Singh, M., Macarisin, D., Sharma, M., and Shelton, D. (2013). Differences in biofilm formation of produce and poultry *Salmonella enterica* isolates and their persistence on spinach plants. *Food Microbiol.* 36, 388-394.

Porto-Fett, A. C., Call, J. E., and Luchansky, J. B. (2008a). Validation of a commercial process for inactivation of *Escherichia coli* O157:H7, *Salmonella* Typhimurium, and *Listeria monocytogenes* on the surface of whole muscle beef jerky. *J. Food Prot.* 71, 918-926.

Porto-Fett, A. C., Hwang, C. A., Call, J. E., Juneja, V. K., Ingham, S. C., Ingham, B. H., and Luchansky, J. B. (2008b). Viability of multi-strain mixtures of *Listeria monocytogenes*, *Salmonella* Typhimurium, or *Escherichia coli* O157:H7 inoculated into the batter or onto the surface of a soudjouk-style fermented semi-dry sausage. *Food Microbiol.* 25, 793-801.

Puttamreddy, S., Carruthers, M. D., Madsen, M. L., and Minion, F. C. (2008). Transcriptome analysis of organisms with food safety relevance. *Foodborne Pathog. Dis.* 5, 517-529.

Robbe-Saule, V., Jaumouillé, V., Prévost, M. C., Guadagnini, S., Talhouarne, C., Mathout, H., Kolb, A., and Norel, F. (2006). Crl activates transcription initiation of RpoS-regulated genes involved in the multicellular behavior of *Salmonella enterica* serovar Typhimurium. *J. Bacteriol.* 188, 3983-3994.

Rodrigues, D., Cerca, N., Teixeira, P., Oliveira, R., Ceri, H., and Azeredo, J. (2011). *Listeria monocytogenes* and *Salmonella enterica* Enteritidis biofilms susceptibility to different disinfectants and stress-response and virulence gene expression of surviving cells. *Microb. Drug Resist.* 17, 181-189.

Römling, U. (2005). Characterization of the rdar morphotype, a multicellular behaviour in Enterobacteriaceae. *Cell. Mol. Life Sci.* 62, 1234-1246.

Römling, U., Sierralta, W. D., Eriksson, K., and Normark, S. (1998). Multicellular and aggregative behaviour of *Salmonella* Typhimurium strains is controlled by mutations in the *agfD* promoter. *Mol. Microbiol.* 28, 249-264.

Ruby, T., McLaughlin, L., Gopinath, S., and Monack, D. (2012). *Salmonella*'s long-term relationship with its host. *FEMS Microbiol Rev.* 36, 600-615.

Runkel, S., Wells, H. C., and Rowley, G. (2013). Living with stress: a lesson from the enteric pathogen *Salmonella enterica*. *Adv. Appl. Microbiol.* 83, 87-144.

Salazar, J. K., Deng, K., Tortorello, M. L., Brandl, M. T., Wang, H., and Zhang, W. (2013). Genes *ycfR*, *sirA* and *yigG* contribute to the surface attachment of *Salmonella enterica* Typhimurium and Saintpaul to fresh produce. *PLoS One* 8(2):e57272.

Scher, K., Romling, U., and Yaron, S. (2005). Effect of heat, acidification, and chlorination on *Salmonella enterica* serovar Typhimurium cells in a biofilm formed at the air-liquid interface. *Appl. Environ. Microbiol.* 71, 1163-1168.

Schonewille, E., Nesse, L. L., Hauck, R., Windhorst, D., Hafez, H. M., and Vestby, L. K. (2012). Biofilm building capacity of *Salmonella enterica* strains from the poultry farm environment. *FEMS Immunol. Med. Microbiol.* 65, 360-365.

Selgas, D., Luisa Marín, M., Pin, C., and Casas, C. (1993). Attachment of bacteria to meat surfaces: a review. *Meat Sci.* 34, 265-273.

Shen, S., and Fang, F. C. (2012). Integrated stress responses in *Salmonella. Int. J. Food Microbiol.* 152, 75-81.

Shi, X., Namvar, A., Kostrzynska, M., Hora, R., and Warriner, K. (2007). Persistence and growth of different *Salmonella* serovars on pre- and postharvest tomatoes. *J. Food Prot.* 70, 2725-2731.

Simm, R., Lusch, A., Kader, A., Andersson, M., and Römling, U. (2007). Role of EAL-containing proteins in multicellular behavior of *Salmonella enterica* serovar Typhimurium. *J. Bacteriol.* 189, 3613-3623.

Simm, R., Morr, M., Kader, A., Nimtz, M., and Römling, U. (2004). GGDEF and EAL domains inversely regulate cyclic di-GMP levels and transition from sessility to motility. *Mol. Microbiol.* 53, 1123-1134.

Smith, J. L., Fratamico, P.M., and Uhlich, G. A. (2009). Molecular mechanisms involved in biofilm formation by food-associated bacteria. *In* P. M. Fratamico, B. A. Annous, & N. W. Gunther (Eds.), Biofilms in the food and beverage industries (pp. 42–98). Oxford, UK: Woodhead Publishing, Ltd.

Sofos, J. N., Kochevar, S. L., Bellinger, G. R., Buege, D. R., Hancock, D. D., Ingham, S. C., Morgan, J. B., Reagan, J. O., and Smith, G. C. (1999). Sources and extent of microbiological contamination of beef carcasses in seven United States slaughtering plants. *J. Food Prot.* 62, 140-145.

Solano, C., García, B., Valle, J., Berasain, C., Ghigo, J. M., Gamazo, C., and Lasa, I. (2002). Genetic analysis of *Salmonella* Enteritidis biofilm formation: critical role of cellulose. *Mol. Microbiol.* 43, 793-808.

Somers, E. B., Schoeni, J. L., and Wong, A. C. (1994). Effect of trisodium phosphate on biofilm and planktonic cells of *Campylobacter jejuni*, *Escherichia coli* O157: H7, *Listeria monocytogenes* and *Salmonella* Typhimurium. *Int. J. Food Microbiol.* 22, 269-276.

Spector, M. P., and Kenyon, W. J. (2012). Resistance and survival strategies of *Salmonella enterica* to environmental stresses. *Food Res. Int.* 45, 455–481.

Steenackers, H., Hermans, K., Vanderleyden, J., and De Keersmaecker, S. C. J., 2012. *Salmonella* biofilms: an overview on occurrence, structure, regulation and eradication. *Food Res. Int.* 45, 502–531.

Stewart, P. S., and Costerton, J. W. (2001). Antibiotic resistance of bacteria in biofilms. *Lancet.* 358, 135-138.

Stewart, P. S., and Franklin, M. J. (2008). Physiological heterogeneity in biofilms. *Nat. Rev. Microbiol.* 6, 199-210.

Stoodley, P., Sauer, K., Davies, D. G., and Costerton, J. W. (2002). Biofilms as complex differentiated communities. *Annu. Rev. Microbiol.* 56, 187-209.

Tabak, M., Scher, K., Hartog, E., Romling, U., Matthews, K. R., Chikindas, M. L., and Yaron, S. (2007). Effect of triclosan on *Salmonella* Typhimurium at different growth stages and in biofilms. *FEMS Microbiol. Lett.* 267, 200-206.

Tamblyn, K. C., Conner, D. E., and Bilgili, S. F. (1997). Utilization of the skin attachment model to determine the antibacterial efficacy of potential carcass treatments. *Poult. Sci.* 76, 1318-1323.

Tan, S. Y., Chew, S. C., Tan, S. Y., Givskov, M., and Yang, L. (2014). Emerging frontiers in detection and control of bacterial biofilms. *Curr. Opin. Biotechnol.* 26, 1-6.

Turnock, L. L., Somers, E. B., Faith, N. G., Czuprynski, C. J., and Lee, W. A. (2002). The effects of prior growth as a biofilm on the virulence of *Salmonella typhimurium* for mice. *Comp. Immunol. Microbiol. Infect. Dis.* 25, 43-48.

Van Houdt, R., and Michiels, C. W. (2010). Biofilm formation and the food industry, a focus on the bacterial outer surface. *J. Appl. Microbiol.* 109, 1117-1131.

Van Puyvelde, S., Steenackers, H. P., and Vanderleyden, J. (2013). Small RNAs regulating biofilm formation and outer membrane homeostasis. *RNA Biol.* 10, 185-191.

Vestby, L. K., Møretrø, T., Balance, S., Langsrud, S., and Nesse, L. L. (2009a). Survival potential of wild type cellulose deficient *Salmonella* from the feed industry. *BMC Vet. Res.* 5:43.

Vestby, L. K., Møretrø, T., Langsrud, S., Heir, E., and Nesse, L. L. (2009b). Biofilm forming abilities of *Salmonella* are correlated with persistence in fish meal- and feed factories. *BMC Vet. Res.* 5:20.

Wan Norhana, M. N., Poole, S. E., Deeth, H. C., and Dykes G. A. (2010). The effects of temperature, chlorine and acids on the survival of *Listeria* and *Salmonella* strains associated with uncooked shrimp carapace and cooked shrimp flesh. *Food Microbiol.* 27, 250-256.

Wang, R., Kalchayanand, N., Schmidt, J. W., and Harhay, D. M. (2013). Mixed biofilm formation by Shiga toxin-producing *Escherichia coli* and *Salmonella enterica* serovar Typhimurium enhanced bacterial resistance to sanitization due to extracellular polymeric substances. *J. Food Prot.* 76, 1513-1522.

Wang, S., Phillippy, A. M., Deng, K., Rui, X., Li, Z., Tortorello, M. L., and Zhang, W. (2010). Transcriptomic responses of *Salmonella enterica* serovars Enteritidis and Typhimurium to chlorine-based oxidative stress. *Appl. Environ. Microbiol.* 76, 5013-5024.

Warriner, K., and Namvar, A. (2010). The tricks learnt by human enteric pathogens from phytopathogens to persist within the plant environment. *Curr. Opin. Biotechnol.* 21, 131-136.

Waterman, S. R., and Small, P. L. (1998). Acid-sensitive enteric pathogens are protected from killing under extremely acidic conditions of pH 2.5 when they are inoculated onto certain solid food sources. *Appl. Environ. Microbiol.* 64, 3882-3886.

White, A. P., Weljie, A. M., Apel, D., Zhang, P., Shaykhutdinov, R., Vogel, H. J., and Surette, M. G. (2010). A global metabolic shift is linked to *Salmonella* multicellular development. *PLoS One* 5(7):e11814.

Xu, H., Lee, H. Y., and Ahn, J. (2010). Growth and virulence properties of biofilm-forming *Salmonella enterica* serovar Typhimurium under different acidic conditions. *Appl. Environ. Microbiol.* 76, 7910-7917.

Zhang, X. S., García-Contreras, R., and Wood, T. K. (2007). YcfR (BhsA) influences *Escherichia coli* biofilm formation through stress response and surface hydrophobicity. *J. Bacteriol.* 189, 3051-3062.

Zheng, J., Allard, S., Reynolds, S., Millner, P., Arce, G., Blodgett, R. J., and Brown, E. W. (2013). Colonization and internalization of *Salmonella enterica* in tomato plants. *Appl. Environ. Microbiol.* 79, 2494-2502.

In: Salmonella
Editor: Christopher B. Hackett

ISBN: 978-1-63463-651-3

Chapter 7

ROLE OF *SALMONELLA* METABOLIC GENES IN THE COLONIZATION OF FOOD ANIMALS

Katherine A. Miller and Timothy R. Hoover*

Department of Microbiology, University of Georgia,
Athens, Georgia, US

ABSTRACT

Salmonella colonize the intestinal tract of a wide variety of animals. Within the intestinal tract, *Salmonella* tend to be scavengers or secondary degraders that are dependent upon nutrients released by the enzymatic action of the more abundant primary degraders which digest intestinal glycoproteins and complex carbohydrates. *Salmonella* are catalytically robust and capable of using a wide variety of carbon and nitrogen sources. While a considerable amount of information is available on *Salmonella* pathogenicity genes and how they facilitate infection, we have only recently begun to appreciate the nutritional requirements of *Salmonella* within the context of the intestinal microbiome. Several recent studies have used –omics driven (i.e., genomics, transcriptomics, proteomics, metabolomics) approaches to investigate possible contributions of metabolic processes to *Salmonella* colonization and virulence. We review here recent research literature on the roles of metabolic genes in *Salmonella* colonization in various host systems including mice, swine, calves and poultry, as well as metabolic genes implicated in the persistence of *Salmonella* on produce. Understanding the nutritional requirements of *Salmonella* in the host intestinal tract is important for identifying potential targets and designing new approaches for reducing growth of *Salmonella* during the food animal production period and thereby mitigating the entry of *Salmonella* into the human food supply.

* Corresponding author: Email: trhoover@uga.edu.

INTRODUCTION

Salmonella are Gram-negative, rod-shaped, flagellated bacteria belonging to the family *Enterobacteriaceae*. There are currently two species of *Salmonella*, *S. bongeri* and *S. enterica*, with *S. enterica* divided into six subspecies. *Salmonella* are classified further by three distinct types of antigens which have been used to distinguish over 2500 serotypes. Thus, the nomenclature for *Salmonella* is complex and includes both the subspecies and serovar designations [1]. For example, the *Salmonella* laboratory strain commonly used for laboratory studies is *S. enterica* subspecies *enterica* serovar Typhimurium, which is shortened to *Salmonella* Typhimurium or *S.* Typhimurium.

Salmonella is a major foodborne pathogen worldwide. A 2011 report estimated that in the United States alone there were approximately 1 million cases of salmonellosis, resulting in 19,000 hospitalizations and 378 deaths [2]. In the United States costs associated with *Salmonella* are estimated to be between 2.3 and 3.6 billion dollars annually [3]. *Salmonella* is usually transmitted to humans through the food chain. Although there have been reports of salmonellosis cases being linked to vegetables and fruits, 95% of salmonellosis cases result from consumption of meat, eggs or poultry [4]. Infection of poultry with most *Salmonella* serovars results in asymptomatic intestinal colonization so reliance on birds appearing to be sick is an ineffective indicator of *Salmonella* colonization. Egg shells can be contaminated with *Salmonella* from infected laying hens, but unless bacterial numbers exceed 10^9 colony forming units per egg no obvious changes in appearance or odor can be observed for eggs which are contaminated with *Salmonella* [5].

One of the ongoing frontiers of *Salmonella* control in food animals is colonization of the intestine. While in some food animals, invasion of the intestinal epithelial cells is essential for colonization, there are many *Salmonella* serotypes that colonize poultry and other food animals which appear to compete quite well within the intestinal bacterial community without the requirement for cellular invasion. A considerable body of work is available on *Salmonella* pathogenicity genes and the mechanisms by which they facilitate infection [6-14]. Until recently, however, there has been relatively little work done on defining nutritional requirements of *Salmonella* that enable them to establish a niche within the intestinal microbiome [10, 11, 15-18]. Studies that have examined how *Salmonella*'s nutritional requirements impact colonization have often focused on global regulators. For example, mutations targeting catabolite repression, specifically *crp* and *cya*, significantly attenuate *Salmonella* without affecting epithelial invasion [16, 19]. Catabolite repression is a regulatory mechanism used by bacteria to prioritize expression of transporters and catabolic enzymes to enable the hierarchical usage of different carbon sources. *Salmonella crp/cya* mutants are reduced in the ability to colonize the chicken cecum, yet we do not understand the mechanism by which *Salmonella* prioritizes the utilization of available carbohydrates enabling it to compete for colonization niches within the intestine.

Successful colonization of the intestinal tract of food animals by *Salmonella* is a complex process that involves not only interactions of *Salmonella* with the host, but also interactions with the gut microbiota. We review here some of the approaches used to identify *Salmonella* genes that appear to have a role in host colonization. We focus on results from these studies that identify *Salmonella* metabolic genes that may be important for colonization of the intestinal tract of various animals. In addition, we examine the research literature on metabolic genes implicated in the persistence of *Salmonella* on produce.

USE OF –OMICS DRIVEN METHODS TO IDENTIFY *SALMONELLA* METABOLIC GENES IMPORTANT FOR COLONIZATION AND VIRULENCE

The application of –omics driven approaches (i.e., genomics, transcriptomics, proteomics and metabolomics) has contributed tremendously to our understanding of metabolic processes that are important for *Salmonella* colonization and virulence. For purposes of this review, we will restrict the definition of metabolic processes to those which are involved in nutrient acquisition (e.g., carbon, nitrogen, sulfur, metal ions) and energy generation. We will highlight some approaches which have been used to identify *Salmonella* genes which have roles in colonization or virulence, and indicate metabolic genes that were identified in these studies as having such roles.

Identification of Salmonella metabolic genes needed for colonization using signature-tagged mutagenesis. Signature-tagged mutagenesis (STM) is a powerful negative selection method used to identify bacterial genes that are potentially important for the successful colonization of a host animal. The method was developed about twenty years ago by David Holden and co-workers to identify mutants in *S.* Typhimurium attenuated for virulence in a murine model of typhoid fever [20]. STM allows for the identification of attenuated mutants from a large pool of mutants by combining transposon-mutagenesis and negative selection with a detection system. Short individual DNA sequences introduced within the transposons (i.e., signature-tags) are used to individually mark the mutants. Pools of mutants with distinct signature-tags are assembled and the mutant pools are subject to a negative selection by inoculating them into an animal host. Mutants that survive the screen are recovered from the animal host (output pool) for detection of the individual signature-tags. Mutants that fail to survive the screen are identified from cultures of the mutant pools used to inoculate the animals (input pool) using the same detection system. The original STM method used hybridization analysis to identify signature-tags, but this detection system has since been supplanted by less laborious methods, including polymerase chain reaction (PCR)-amplification, real-time PCR and DNA microarrays [21-23].

Roles for genes in colonization which have been identified by STM need to be verified by reconstructing the relevant mutations in the original parental strain and then testing the resulting mutants for colonization defects. A drawback of STM is that it is not possible to confirm colonization phenotypes for all (or even most) of the mutant alleles identified from screens that yield large numbers of potential colonization mutants. The colonization phenotype of a reconstructed mutant strain can be examined by inoculating the host animal with a pure culture of the strain or by using a competitive index (CI) assay in which a mixture of the mutant and parental strain is used to inoculate the host animal. In the CI assay, a mutant strain is typically mixed in a 1:1 ratio with its parental strain and the resulting mixture is used to inoculate the host animal. Bacteria recovered from the host post-inoculation are analyzed to determine the ratio of mutant to wild type. Mutations which decrease fitness for colonization result in the recovery of substantially fewer mutants compared to wild type.

Several studies have used STM to identify *Salmonella* genes which have potential roles in colonization or virulence in a variety of animal hosts, including mice, chicks, pigs and calves [8, 14, 20, 24-35]. While most of the work in these studies focused on virulence genes within *Salmonella* pathogenicity islands (SPI), in some cases metabolic genes were identified as having potential roles in colonization. Two of the most comprehensive STM-based studies to identify *Salmonella* genes important for colonization of food animals were done by Morgan et al. [14] and Carnell et al. [8]. Morgan and co-workers identified approximately 150 *S.* Typhimurium genes that appeared to be important for colonization of the intestinal tracts of calves, 2-week-old chicks, or both animal species [14]; while Carnell and colleagues identified 95 S. Typhimurium genes which appear to be needed for intestinal colonization of the pig [8]. Several of the genes identified in these studies are involved in carbohydrate catabolism, sulfur metabolism, amino acid degradation or nutrient transport (Table 1), although the colonization phenotypes were not confirmed for any of these mutants.

Given the caveat that not all of the genes listed in Table 1 may ultimately prove to be important for colonization, a couple of observations are notable. First, some genes appear to be specifically required in one animal host but not others. For example, the carnitine/γ-butyrobetaine antiporter CaiT is required for efficient intestinal colonization of the chick, but not for colonization of the calf or pig. L-carnitine is a commonly found trimethylammonium compound which *Salmonella* and other members of the *Enterobacteriaceae* are able to reduce to γ-butyrobetaine under anaerobic conditions [36]. The reduction of L-carnitine stimulates fermentative growth of *S.* Typhimurium by allowing the bacterium to regenerate NAD^+ [36]. Given that the intestinal tract is an anaerobic environment, the ability of *S.* Typhimurium to utilize L-carnitine might be expected to be an important colonization factor. That L-carnitine utilization does not appear to play an important role in *S.* Typhimurium colonization of the pig or calf suggests that other electron acceptors are available to *S.* Typhimurium for fermentative growth or anaerobic respiration in these host animals.

Table 1. *S.* Typhimurium metabolic genes identified from STM studies as having possible roles in intestinal colonization of the chick, calf or pig

Classification	Gene name	Known or predicted function	[a]chick	[b]calf	[c]pig
Carbon compound degradation / central metabolism	*kduD*	2-deoxy-D-gluconate 3-	- (50)	wt	wt
	STM3600	dehydrogenase		wt	wt
	STM3793	putative sugar kinase	- (50)	-	wt
	rbsK	putative carbohydrate kinase	wt	wt	wt
	citC	ribokinase	- (33)	-	-
	celF	citrate lyase synthetase	- (50)	n.d.	-
	kduI	cellobiose-6-phosphate hydrolase	n.d.	n.d.	-
	pflA	5-keto-4-deoxyuronate isomerase	n.d.	n.d	-
		pyruvate formate lyase activating enzyme I	n.d.		
Sulfur metabolism	STM0036	putative arylsulfatase regulator	- (100)	wt	-
	STM0084	putative sulfatase	- (100)	wt	wt
Amino acid degradation	STM2196	putative L-serine dehydratase	- (100)	wt	wt

Classification	Gene name	Known or predicted function	[a]chick	[b]calf	[c]pig
Transport/ binding	*caiT*	carnitine transport	- (100)	wt	wt
	STM0328	putative permease	- (100)	-	-
	sfbB	ABC transporter	- (100)	wt	wt
	fepA	ferric enterobactin receptor	- (50)	-	-
	entF	enterobactin synthetase	- (100)	-	wt
	STM0765	putative cation transporter	- (100)	-	-
	potH	putrescine transporter	wt	-	wt
	focA	probable formate transporter	- (100)	wt	wt
	hpaX	4-hydroxyphenylacetate permease	- (100)	-	wt
	chaA	sodium-calcium/proton antiporter	- (50)	wt	wt
	hisM	histidine transport system permease	- (50)	-	wt
	gabP	gamma-aminobutyrate permease	- (50)	wt	wt
	yneA	ABC transporter	- (50)	-	wt
	cadB	cadaverine-lysine antiporter	n.d.	n.d.	-
	STM2574	putative permease	n.d.	n.d.	-

[a] Data from Morgan et al. [14]. The '-' denotes mutant showing attenuated colonization; 'wt' indicates wild-type colonization phenotype for the mutant; and 'n.d.' indicates the colonization phenotype for the mutant was not determined. Number in parentheses indicates the percent of birds in which the mutant was attenuated.

[b] Data from Morgan et al. [14].

[c] Data from Carnell et al. [8].

Another notable observation regarding the data presented in Table 1 is that many of the metabolic genes appear to be required for efficient colonization of a range of animal hosts. For example, the ferric enterobactin receptor FepA was found to be an important colonization factor in all three host animals. During iron starvation, bacteria often utilize high-affinity iron chelators known as siderophores to scavenge Fe^{2+} from the surrounding medium. Enterobactin is a siderophore produced by *S.* Typhimurium which it uses for iron acquisition. Enterobactin synthetase (EntF) appears to be important for intestinal colonization of the chick and calf, but not the pig intestinal tract. This observation does not necessarily indicate that enterobactin is not required for iron acquisition by *S.* Typhimurium in pigs as the *entF* mutant could have utilized enterobactin produced by other *S.* Typhimurium mutants or other bacteria in the pig intestinal microbiota.

Despite the wealth of available information on *Salmonella* physiology, functions for several of the genes listed in Table 1 are unknown. Identifying functions for these genes will lead to a better understanding of *Salmonella* physiology. Moreover, it seems reasonable to assume that *Salmonella* encounters a different array of nutrients in the various host animals which it colonizes, and identifying activities for these genes of unknown function will provide valuable information on the variety of nutrients *Salmonella* encounters and is able to utilize within various host animals.

Use of transposon-directed insertion-site sequencing to identify Salmonella metabolic genes needed for colonization. While STM has proven to be a powerful method for identifying genes with potential roles in colonization, the task of assigning genotypes with colonization phenotypes is somewhat time consuming. Transposon-directed insertion-site sequencing (TraDIS) circumvents this problem by simultaneously identifying the location of the transposon within a mutant and the relative fitness of the mutant. As with STM, a negative selection method is used in TraDIS to identify bacterial genes important for colonization by inoculating pools of transposon-insertion mutants in a host animal and then recovering colonization-proficient bacteria from the animal. The locations of transposons in the mutants are identified by parallel sequencing of transposon-flanking regions using DNA isolated from the input and output pools. DNA sequences from transposon insertions that result in colonization attenuation are underrepresented in the output pool relative to the input pool. Chaudhuri and co-workers used the TraDIS method to screen pools of random insertion mutants of *S.* Typhimurium in chicks, pigs, and calves [37]. Over 7,700 transposon-insertion mutants were mapped and assigned colonization phenotypes to 2,715 different genes in the study. Fitness scores were defined as the log_2-fold change in the number of sequence reads obtained across the boundaries of each transposon insertion between the input and output pools. The researchers estimated *P* values using biological replicates (duplicates or triplicates of each mutant pool), and defined attenuated mutants as those with a negative fitness score and $P \leq 0.05$ [37]. Colonization phenotypes for twelve of the mutants identified from the TraDIS assay were tested by creating null mutants with defined deletions and using a CI assay in chicks to verify the predicted colonization defect. At day 4 post-inoculation, which corresponded to the time at which mutants were recovered for the TraDIS analysis, statistically significant differences in the ratio of mutant to wild-type strains recovered from the chicks compared to the ratio in the inocula were observed for eight of the null mutants, while significant differences were detected at later time points for two additional mutants [37]. These data suggest that the reliability of the TraDIS assay in predicting colonization phenotypes of individual mutants within a large mutant pool is fairly good. Further evidence supporting the reliability of TraDIS assay in predicting colonization phenotypes was the observation that multiple independent mutations for a given gene or pathway often resulted in attenuation (see Table 2 and reference [37]).

A somewhat unexpected result from the study by Chaudhuri et al. [37] was the seemingly high number of transposon-insertion mutants which were attenuated. Of the 7,700 transposon-insertion mutants that were screened and mapped in the TraDIS assays in the food-producing animals, 25%, 28% and 36% were identified as being attenuated in chicks, pigs and calves, respectively. Most of the genes identified as being potentially involved in colonization were required in all three animals, but smaller subsets of genes appeared to be host-specific. Many of the transposon insertions were in SPI genes or other genes which were known previously to be important for colonization. Many of the mutations, however, were in genes involved in nutrient acquisition or energy generation and had not been shown previously to be important for colonization. Examples of genes involved in catabolism, nutrient acquisition or energy generation that appear to be highly relevant for colonization of chicks, pigs or calves are listed in Table 2. To generate the gene list, we picked a set of 30 mutants (within unique genes) with the lowest *P*-values from the statistical analysis of the TraDIS data for each animal host, reasoning that these should be some of the most highly attenuated mutants.

As seen in Table 2, mutations within a given gene did not always result in the same colonization phenotype in the TraDIS assay. These variances might result from the location or orientation of the transposon within the gene. The location of the transposon could affect the degree to which the insertion exerts polar effects on downstream genes. Alternatively, some transposon insertions within a given gene could result in the complete loss of activity of the product of the gene while other insertions might result in the expression of truncated proteins that retain most or all of their function. It is possible that some such truncated proteins have unregulated activities that are deleterious.

From the information presented in Table 2, we can make some generalizations which have been noted already. For instance, while most of the genes listed in Table 2 appear to be important for efficient colonization in all three of the food-producing animals there do appear to be some genes which are host specific. Two such examples of host-specific genes are *pgi* and *pfkA*, which encode the glycolytic enzymes glucose-6-phosphate isomerase (catalyzes the reversible isomerization of glucose-6-phosphate to fructose-6-phosphate) and 6-phosphofructokinase (catalyzes the irreversible phosphorylation of fructose-6-phosphate to fructose-1,6-bisphosphate). These glycolytic enzymes appear to be important for efficient colonization in the chick (and pig to a lesser degree), but not in the calf. Since glucose levels are presumably quite low in the chick cecum [12], *S.* Typhimurium most likely needs glycolysis for chick colonization in order to generate precursor metabolites for biosynthesis rather than energy generation. Consistent with this hypothesis, mutants with transposon-insertions in genes required for the biosynthesis of branched chain amino acids (isoleucine, leucine and valine) or histidine tended to be more attenuated for chick colonization than calf colonization in the TraDIS assays [37]. Biosynthesis of these amino acids requires precursor metabolites derived from glycolysis, pyruvate for branched chain amino acids and fructose-6-phosphate for histidine. This suggests that branched chain amino acids and histidine are limiting in the chick cecum (but not the calf intestinal tract), and that *Salmonella* must be able to synthesize these amino acids to colonize the chick cecum. Such differences in the availability of these amino acids could be attributed to differences in the hosts' diets, physiologies or microbiota.

As noted for the data presented in Table 1, functions are not known for many of the genes listed in Table 2. There are a couple of possible reasons for failure to assign known or predicted functions to these genes. The most obvious reason is that the products of the genes have not been analyzed biochemically. A second reason is that the genes were not carefully annotated in the genome sequence databases. Examples of potential gene assignments which have gone unnoticed in automated gene annotations are found in the genes *gfrC* (SL1344_4468), *gfrE* (SL1344_4470) and *grfF* (SL1344_4471) in Table 2. In 2005, Wiame and co-workers reported that the *Enterococcus faecium grfABCDEF* operon encode a novel phosphotransferase system (PTS) permease and two deglycases involved in fructoselysine and glucoselysine catabolism [38]. These researchers identified a homologous operon in *S.* Typhimurium and predicted that it was similarly involved in fructoselysine and glucoselyine catabolism [38]. We constructed null mutants for homologs of *gfrAB*, *gfrE* and *gfrF* in *S.* Typhimurium and confirmed that these genes are indeed required for utilization of fructoselysine and glucoselyine (K. A. Miller, unpublished data). Fructoselysine is formed by the reaction of glucose with the ε-amine of lysine followed by a spontaneous isomerization reaction (Amadori rearrangement) [39]. Glucoselysine is similarly formed from the reaction of fructose with the ε-amine of lysine, but the resulting product can undergo a couple of possible rearrangements, one of which is the 2-amino-aldose derivative with either a glucosamine (i.e., glucoselysine) or a mannosamine (i.e., mannoselysine) configuration [40]. The spontaneous reactions of amines with glucose or fructose are designated as glycation and fructation, respectively. We postulate that *Salmonella* encounters fructoselysine and glucoselysine in the intestinal tracts of food-producing animals when proteins modified by glycation or fructation are degraded by proteases derived from the host or primary degraders in the gut microbiota. These modified proteins could come from feed, the host or members of the microbiota.

Table 2. Examples of *S.* Typhimurium metabolic genes that appear to have roles in intestinal colonization of the chick, pig or calf

Classification	Gene name	Known or predicted function	[d]Number of attenuated mutants versus total number of mutants		
			chick	pig	calf
Carbohydrate metabolism	[a]*mtlD*	mannitol-1-phosphate dehydrogenase	5/5	5/5	4/5
	[a]*tktA*	transketolase	5/5	5/5	5/5
	[a]*pgi*	glucose-6-phosphate isomerase	2/2	1/2	0/2
	[a]*pfkA*	6-phosphofructokinase	2/2	2/2	0/2
	[a,c]*gfrE*	glucoselysine-6-phosphate deglycase	2/3	2/3	2/2
	[a,b]*aceF*	E2 component of pyruvate dehydrogenase	2/2	2/2	2/2
	[a]SL1344_3746	hypothetical fructose-1,6-bisphosphate aldolase	2/2	2/2	2/2
	[a]*rpe*	ribulose-phosphate-3-epimerase	3/3	3/3	3/3
	[a]*pflB*	formate acetyltransferase 1	5/6	2/6	2/6
	[a]SL1344_3565	hypothetical carbohydrate kinase	3/4	3/3	4/4
	[b]*nagA*	N-acetylglucosamine-6-phosphate deacetylase	2/2	2/2	2/2
	[b,c]SL1344_3106	putative xylanase/chitin deacetylase	3/6	4/5	4/4
	[b]*eda*	KHG/KDPG aldolase	1/1	1/1	0/0
	[b]SL1344_3059	putative mannitol dehydrogenase	1/3	1/3	1/3
	[b]*gfrF*	fructoselysine-6-phosphate deglycase	5/6	5/6	5/5
	[c]*idnD*	L-idonate 5-dehydrogenase	3/4	3/4	4/4
	[c]SL1344_2738	hypothetical hexulose 6- phosphate synthase	3/3	3/3	3/3
	[c]*iolG2*	myo-inositol 2-dehydrogenase	1/2	1/1	2/2
	[c]SL1344_2242	putative dehydratase	4/4	3/3	4/4
	[c]SL1344_1471	putative zinc-binding dehydrogenase	2/2	2/2	2/2
	[c]*iolG1*	myo-inositol 1-dehydrogenase	1/8	2/6	4/6
Nitrogen metabolism	[a,b]SL1344_3563	L-asparaginase	3/7	5/6	3/7
	[a,b]*tdcB*	catabolic threonine dehydratase	6/6	6/6	4/4
	[b]*tdcD*	proprionate kinase	1/3	2/3	3/3
Sulfur metabolism	[a]*trmE*	thiophene/furan oxidation	3/3	3/3	3/3
	[a]SL1344_0084	hypothetical sulfatase	6/7	7/7	7/7
	[a]SL1344_3959	possible sulfatase	9/13	11/12	11/12
	[a]*yejM*	hypothetical sulfatase	2/2	2/2	2/2
	[a]*yheM*	putative oxidation of intracellular sulfur	2/2	2/2	2/2
	[a]SL1344_0033	possible sulfatase	2/2	1/1	2/2
	[b]*assT*	probable arylsulfate sulfotransferase	1/4	2/2	3/3

Classification	Gene name	Known or predicted function	[d]Number of attenuated mutants versus total number of mutants		
			chick	pig	calf
	[c]SL1344_4047	hypothetical arylsulfate sulfotransferase	5/9	7/9	7/8
	[c]*asrA*	anaerobic sulfite reductase subunit A	2/3	2/3	2/3
Transport/ binding	[a]*pstC*	phosphate transport	3/3	3/3	2/3
	[a]*pstB*	phosphate transport	3/3	3/3	2/2
	[a,c]*yjeM*	putative amino acid transporter	4/8	3/7	6/8
	[a,c]*ycaM*	putative amino acid transporter	4/5	4/4	5/5
	[a]SL1344_3748	IIC component of sugar-PTS permease	5/7	6/6	6/6
	[a,c]SL1344_3799	hypothetical transporter	8/12	10/12	9/11
	[a]*znuB*	high-affinity zinc transporter, membrane protein	1/1	1/1	1/1
	[a,c]*mgtC*	magnesium transporter, ATPase protein C	1/2	2/2	2/2
	[a]*focA*	probable formate transporter	3/3	2/3	2/3
	[a]*ydeZ*	putative sugar transport protein	3/6	5/6	6/6
	[b]*gfrC*	IIC component of fructose- /glucoselysine PTS permease	1/1	1/1	1/1
	[b]*yrbF*	putative ABC transporter, ATP-binding protein	1/1	1/1	1/1
	[b,c]*ptsI*	phosphotransferase system (PTS), enzyme I	3/3	3/3	3/3
	[b]SL1344_0860	possible transport protein	1/2	1/2	2/2
	[b]*pstA*	phosphate transport system permease protein	3/3	2/3	2/3
	[b]*iroN*	TonB-dependent siderophore receptor protein	4/7	5/7	3/6
	[b]*ybbL*	putative ABC transporter, ATP-binding protein	1/1	1/1	1/1
	[b]*fepG*	ferric enterobactin transport protein	1/1	1/1	1/1
	[b]*fepC*	ferric enterobactin transport ATP-binding protein	1/1	1/1	1/1
	[b]SL1344_3750	EIIA component of PTS permease	4/4	4/4	3/3
	[b]SL1344_2536	putative permease	1/1	1/1	1/1
	[b]*mglB*	D-galactose-binding periplasmic protein precursor	1/4	2/3	1/2
	[c]SL1344_4000	oligogalacturonate-specific porin(KdgM) family member	1/2	1/2	2/2
	[c]*yiaM*	TRAP-type transporter, small permease protein	1/1	1/1	1/1
	[c]*yhjV*	hypothetical amino acid permease	5/9	7/7	9/9
	[c]SL1344_4001	TRAP-type transporter, large permease component	2/4	1/2	2/4
	[c]SL1344_0940	hypothetical ion:amino acid symporter	2/2	2/2	2/2
	[c]SL1344_3100	possible amino acid transport protein	2/6	2/6	5/6
	[c]*ybbY*	putative purine permease	3/4	4/4	4/4
	[c]SL1344_3000	magnesium transporter, ATPase protein C	6/12	5/9	9/12

Table 2. (Continued)

Classification	Gene name	Known or predicted function	[d]Number of attenuated mutants versus total number of mutants		
			chick	pig	calf
	[c]*proV*	glycine betaine/l-proline transport ATP-binding protein	1/5	5/5	5/5
	[c]SL1344_3758	probable permease	1/3	1/3	2/3
Regulatory protein for catabolic pathway(s) or nutrient acquisition	[a,b,c]*fruR*	fructose repressor	4/4	4/4	4/4
	[b]*crp*	cyclic AMP receptor protein	2/2	2/2	1/1
	[b]*cyaA*	adenylate cyclase	2/2	2/2	2/2
	[b]*phoU*	phosphate transport system regulatory protein	2/2	2/2	1/1
	[b]*citC*	activates citrate lyase by acetylation	0/4	1/2	4/4
Energy generation	[a]*hybD*	hydrogenase-2 component	2/2	0/2	0/1
	[b]*nuoC*	NADH dehydrogenase I, chain C	5/7	5/5	5/5
	[b]*nuoA*	NADH dehydrogenase I, chain A	2/2	1/1	2/2
	[b]*atpD*	ATP synthase beta subunit	3/3	1/3	3/3
	[c]*cydA*	cytochrome d ubiquinol oxidase subunit I	1/1	1/1	1/1
	[c]*atpA*	ATP synthase alpha subunit	5/5	3/3	4/4
	[c]*nuoJ*	NADH dehydrogenase I, chain J	0/2	2/2	2/2
	[c]*hypO*	hydrogenase-2 small chain protein	4/5	1/5	2/5

[a] Transposon insertions in *S.* Typhimurium SL1344 metabolic genes that displayed the most statistically significant colonization defect (i.e., lowest E value) in the chick colonization model. The top 30 genes are indicated.

[b] Transposon insertions in *S.* Typhimurium SL1344 metabolic genes that displayed the most statistically significant colonization defect (i.e., lowest E value) in the pig colonization model. The top 30 genes are indicated.

[c] Transposon insertions in *S.* Typhimurium SL1344 metabolic genes that displayed the most statistically significant colonization defect (i.e., lowest E value) in the calf colonization model. The top 30 genes are indicated.

[d] Number of different insertion mutations in gene which displayed a statistically significant colonization defect (E<0.05) versus total number of insertion mutations for gene. Data presented in table are from Chaudhuri et al. [37].

Once a metabolic gene has been identified as having a potential role in colonization using a transposon-based screening method such as STM or TraDIS, one still has the tasks of first verifying that it is disruption of the gene containing the transposon insertion which is responsible for the colonization phenotype, and second, determining why inactivation of the gene interferes with colonization. With regard to verifying genotype-phenotype assignments, it is important to rule out polar effects on genes located downstream of the transposon insertion as being responsible for the colonization defect. This is typically done by constructing an unmarked deletion of the gene using the lambda Red recombination (referred to as recombineering) system [41] and then determining the colonization phenotype of the resulting null mutant. With regard to understanding of how a mutation affects colonization, there are a couple of scenarios which could account for how loss of a metabolic gene interferes with colonization. First, the mutation might directly block a pathway or process that is necessary for growth of the bacterium in the animal host. The colonization defects of strains with mutations in the genes listed under the classification of 'Energy generation' in Table 2 are most likely explained by this scenario. A second scenario is the mutation prevents the synthesis of an essential building block (e.g., amino acid, nucleotide, sugar) by preventing formation of a precursor metabolite. As discussed previously, we postulate that this scenario may account for the colonization defects for *S.* Typhimurium strains with mutations in *pgi* and *pfkA*. A third scenario is the mutation results in an accumulation of a toxic metabolic intermediate. Precedence for this hypothesis comes from the observation that elevated levels of sugar phosphates in *E. coli* can lead to the cessation of cell growth or even cell death [42-44]. Disruption of *mtlD*, which encodes mannitol-1-phosphate-5-dehydrogenase, resulted in attenuation for colonization of *S.* Typhimurium in the chick, pig and calf in the TraDIS assay [37]. We postulate that the *mtlD* mutants are attenuated due to the accumulation of mannitol-1-phosphate since strains with transpospon insertions in *mtlA*, which encodes a mannitol-specific PTS permease, display wild-type levels of colonization in all three animal hosts. MtlA phosphorylates mannitol as it is transported across the membrane, and MtlD subsequently converts mannitol-1-phosphate to fructose-6-phosphate. Thus, knocking out *mtlD* would be expected to result in an intracellular accumulation of mannitol-1-phosphate when mannitol is available extracellularly.

Transcriptome analysis to identify Salmonella metabolic genes involved in colonization. Transcriptomics, also referred to as expression profiling, examines the expression levels of mRNAs in a cell population using high-throughput methods such as DNA microarrays or next-generation sequencing technology (RNA-Seq). Harvey and co-workers used a whole-genome DNA microarray to examine gene expression of *S.* Typhimurium harvested from chick ceca, comparing the expression pattern with that of bacteria grown in broth culture [12]. The researchers identified 282 *S.* Typhimurium genes which were up-regulated >2-fold in bacteria recovered from the luminal contents of the chick ceca compared to bacteria grown in broth culture [12]. Genes required for utilization of propionate (*prp* operon), ethanolamine (*eut* operon), 1,2-propanediol (*pdu* operon), melibiose (*melA*) and L-ascorbate (*yiaM*, *yiaN*, *lyxK* and *sbgH*) were all up-regulated in *S.* Typhimurium from the chick ceca, suggesting that these are carbohydrates that are utilized by *Salmonella* in the chick cecum. Other metabolic genes which were up-regulated in *S.* Typhimurium isolated from the chick ceca included genes required for the transport or catabolism of putrescine (*potFG*) and arginine (*artJ*, *speA*, *adi*, *argA*), again suggesting that S. Typhimurium encounters and utilizes these compounds in the chick ceca.

Harvey and co-workers' identification of propionate, enthanolamine and 1,2-propanediol as carbon sources for *S.* Typhimurium in the chick cecum is an important observation and is consistent with results from similar studies in other host animals. Their results, however, represent only a brief snapshot of the metabolic processes *Salmonella* uses for efficient and long-term colonization of the intestinal tract. In their study, Harvey and co-workers inoculated chicks which were less than one-day old and sacrificed the chicks 16 hours post-inoculation. In addition, the chicks were given sterile water but were not fed during this period, with their only nourishment coming from the unabsorbed yolk sac. Under such conditions, the chicks would not have developed the normal intestinal microflora and the bacteria within the intestinal tract would not receive the variety of nutrients derived from the feed. Thus, based on the results from the study by Harvey et al. [12] it is impossible to rule out roles for metabolic pathways in *S.* Typhimurium that might be needed for colonization in the presence of the normal intestinal microflora or as the birds mature.

Jay Hinton and co-workers recently published an extensive analysis of the *S.* Typhimurium transcriptome using RNA-Seq based technology [45]. While this study did not examine the *S.* Typhimurium transcriptional profile in an animal host, the researchers did use 22 distinct growth conditions which mimicked environmental conditions relevant to infection. Some of the conditions were anaerobic growth, growth at 25°C, and shock to a variety of insults, including acid, bile, NaCl, low iron, nitric oxide, and peroxide. As expected, the various conditions resulted in characteristic transcription profiles, and 86% of all of the *S.* Typhimurium genes were induced under one or more of the conditions [45]. The study is a valuable resource for the *Salmonella* research community as it provides a comprehensive report on how infection-relevant environment conditions affect global transcription in *S.* Typhimurium. In addition, the Hinton lab has provided a web-based site where the expression profile data can be easily browsed (http://bioinf.gen.tcd.ie/cgi-bin/salcom.pl).

Identification of Salmonella metabolic genes involved in colonization using proteomic approaches. Proteomics is the large-scale study of proteins and it generally involves the identification, and sometimes the quantification, of the entire set of proteins produced by an organism.

A common proteomic procedure is liquid chromatograph-tandem mass spectrometry (LC-MS/MS). In LC-MS/MS, protein samples from an organism are digested with trypsin. The resulting peptides are separated by liquid chromatography and then analyzed by tandem mass spectrometry to identify the peptides based on their predicted atomic masses. We are unaware of any reports of proteomic approaches used to examine the *Salmonella* proteome for bacteria isolated from the intestinal tract of any host animal. Steeb and co-workers [46], however, used proteomics to examine the *Salmonella* proteome in a mouse systemic infection model that mimics human enteric fever. For these studies, the researchers used a cell sorter to isolate *S.* Typhimurium from infected mouse spleen. The *S.* Typhimurium strain used to infect the mice expressed a green fluorescent protein which facilitated the sorting.

The researchers were able to estimate copy numbers for 477 metabolic enzymes, which included 102 transporters and enzymes involved in the 24 pathways for the catabolism of various carbohydrates, amino acids, nucleosides and lipids [46]. To determine which of these pathways were important for intracellular growth and survival of *Salmonella* in host tissue, Steeb and co-workers used a CI assay to analyze a panel of mutants disrupted in the various pathways. Transport proteins were preferentially targeted for deletion rather than enzymes in catabolic pathways to avoid the potential for the accumulation of toxic metabolic intermediates.

The researchers found that *Salmonella* colonization of the mouse spleen was dependent on the ability of the bacterium to utilize glycerol, fatty acids, N-acetylglucosamine, gluconate, glucose, lactate and arginine [46]. These findings indicate that in systemic infections, *Salmonella* is dependent on a complex diet with a variety of nutrients which appear to be available in only scarce amounts.

ACQUISITION OF DIVALENT METAL IONS BY *SALMONELLA* IN THE INTESTINAL TRACT

In addition to –omics driven approaches to identify genes with potential roles in host colonization, researchers have often examined how inactivation of specific metabolic processes or genes impacts colonization. Researchers have used this later approach to determine potential roles for *Salmonella* metal acquisition in host colonization. To establish a niche within the gut microbial community, *Salmonella* must be able to scavenge essential metal ions from the surrounding environment. S. Typhimurium uses a variety of specialized transporters to acquire essential metals in the metal-limited environment of the gut.

One of the most important metals that *Salmonella* requires is iron. Iron is limited in the gut because of sequestration of iron by host proteins like heme, ferretin, transferrin, and lactoferrin [47]. All *Enterbacteriaceae,* including *Salmonella* and commensal *Eschericia coli,* secrete enterochelin, a siderophore which allows the bacteria to scavenge iron in a non-inflamed host [48]. During an inflammatory response induced by *Salmonella* during infection, however, the host secretes lipocalin-2, an antimicrobial peptide that sequesters ferric enterochelin which can limit growth of bacteria that only use enterochelin for iron acquisition [49, 50]. To circumvent the iron limitation during inflammation, *Salmonella* produces salmochelin, a C-glycosylated derivative of enterochelin that is too large to be bound by lipocalin-2 [51, 52]. Thus, salmochelin enhances *Salmonella* colonization in the inflamed intestine [53]. *E. coli* Nissle 1917 is a commensal bacterium that can establish a persistent colonization of the murine intestinal tract and has been used to treat acute enteritis [54]. Interestingly, the *E. coli* Nissle 1917 genome encodes for multiple iron uptake systems including one for salmochelin [55]. This probiotic bacterium reduces *S.* Typhimurium's ability to colonize the intestinal tract by competing for iron [56]. When *Escherichia coli* strain Nissle 1917 is co-inoculated with *S.* Typhimurium, it is able to outcompete and reduce *S.* Typhimurium colonization in the mouse models of acute colitis and chronic persistent infection [56].

Salmonella induces inflammation through its interactions with host cells including epithelial cells and antigen-presenting cells such as macrophages and dendritic cells. Antigen-presenting cells infected by *S.* Typhimurium secrete cytokines that induce a tissue response leading to an influx of neutrophils to the gut mucosa. Neutrophil protein calprotectin is induced during infection with *S.* Typhimurium. Calprotectin, which is the most abundant antimicrobial protein secreted by neutrophils, chelates zinc and inhibits microbial growth by limiting zinc availability [57]. Although calprotectin has antimicrobial properties against many bacteria, it does not inhibit growth of *S.* Typhimurium as the bacterium is able to acquire zinc by expressing a high-affinity zinc transporter (ZnuABC). Mutants that lack ZnuABC are impaired in colonization of the intestinal tract suggesting that the zinc transport system is required for *Salmonella* to compete effectively with the resident microbiota [58].

Salmonella Utilizes Liberated Host Sugars to Promote Expansion

The gut microbiota is made up of primary and secondary degraders like Bacteroides and Firmicutes [59]. The primary degraders break down polysaccharides, oligosaccharides and glycoproteins ingested by the host as well as host mucin to release sugars which they, as well as other bacteria in the gut, can utilize. During normal conditions, where the host is not challenged by a pathogen or treated with antibiotics, most of the nutrients that are released are utilized by the normal microflora, making it difficult for an invading pathogen to find the nutrients it needs to successfully colonize the host. Treatment of the host with antibiotics disrupts the normal intestinal microflora which allows pathogens such as *Salmonella* to establish a niche within the host. Sialic acid and fucose commonly occur as terminal sugars of the oligosaccharides attached to mucin. Although *Salmonella* lacks extracellular enzymes for removing sialic acid and fucose from host glycoproteins, it does possess genes encoding enzymes that enable it to utilize sialic acid (*nan* genes) and fucose (*fuc* genes) freed by primary degraders.

Ng and co-workers [60] explored the interaction that *S.* Typhimurium has with a representative member of the intestinal microflora and primary degrader, *Bacteroides thetaiotaomicron. B. thetaiotaomicron* is a common colonizer of the human gut which produces extracellular enzymes to degrade both dietary and mucin-derived polysaccharides. Ng et al. [60] observed that levels of free sialic acid were increased in mice inoculated with *B. thetaiotaomicron* 6-fold compared to germ-free mice or mice that were colonized with a sialidase-deficient mutant of *B. thetaiotaomicron. B. thetaiotaomicron* does not use the liberated sialic acid, but does release and utilize the sugars that are interior to the terminal sialic acid. In a mouse colitis model, *S.* Typhimurium *nan* and *fuc* genes were shown to be upregulated in a *B. thetaiotaomicron*-dependent manner [60]. Removing the pathways for fucose and sialic acid utilization reduced the competitiveness of *S.* Typhimurium in mice in the presence of *B. thetaiotaomicron* showing that these systems are important for *Salmonella* to grow during infection.

To determine the role of sialic acid in mice containing a complex microflora, Ng et al. treated conventional mice with antibiotics. Before antibiotic treatment, the levels of free sialic acid were low suggesting that the sialic acid that is present is quickly used up by the resident bacteria. After antibiotic treatment, however, there was a spike in the amount of free sialic acid, presumably due to the disruption of the microbiota [60]. It should be noted that high doses of streptomycin treatment results in a mild inflammatory response in mice [61], and *Salmonella* presumably takes advantage of both the disruption of the microflora as well as the host inflammatory response.

Host Inflammation Can Promote Growth of *Salmonella*

Studies of *S.* Typhimurium in the streptomycin-treated mouse colitis model have shown that certain virulence factors induce inflammation resulting in the production of compounds that can serve as terminal electron receptors to support anaerobic respiration in *Salmonella* during infection [61]. In the 1980s, a multidrug-resistent *S.* Typhimurium strain that contained a phage encoding type III effector SopE caused an epidemic among cattle and humans [62, 63].

SopE induces intestinal inflammation in the host [64]. *S.* Typhimurium strains that possess the phage-encoded SopE stimulate the host to produce nitric oxide synthetase (iNOS) in the intestine. Nitric oxide generated under these conditions can be converted to nitrate which is a desirable electron acceptor for *Salmonella* and results in an outgrowth of *Salmonella* in the intestinal lumen [65]. In addition to stimulating nitric oxide production in the host, SopE also suppresses the expression of *S.* Typhimurium genes involved in the utilization of less favorable electron acceptors such as tetrathionate [65].

Tetrathionate utilization is common in *Salmonella* and is used to differentiate it from other *Enterobacteriaceae*. The *ttrSR ttrBCA* gene cluster in *Salmonella* is required for utilization of tetrathionate as a terminal electron acceptor. Colonic bacteria produce high amounts of hydrogen sulfide (H_2S), which is extremely toxic. To circumvent this toxicity, the cecal mucosa converts hydrogen sulfide to thiosulfate ($S_2O_3^{2-}$) [66]. Thiosulfate cannot be used as an electron receptor by the *ttrSR ttrBCA* gene cluster [67]. However, thiosulfate can be oxidized during intestinal inflammation to tetrathionate ($S_4O_6^{2-}$) [68], which can be used as a respiratory electron acceptor for *Salmonella*. Under the anaerobic conditions of the intestinal lumen, bacteria compete for high-energy nutrients for fermentation. Using tetrathionate as an electron acceptor for anaerobic respiration allows *Salmonella* to use products that can only be respired, which provides the bacterium with a growth advantage. Using the mouse colitis model, Winter et al. showed that a *ttrA* mutant, which is defective in tetrathionate utilization, is attenuated for colonization of the mouse intestinal tract [68]. By analyzing the microbiota composition, it was shown that the wild-type strain, but not the *ttrA* mutant, was able to out-compete the resident microbiota in the cecum [68]. Ethanolamine is a major non-fermentable carbon source found in the intestinal tract. *S.* Typhimurium's ability to respire ethanolamine in the presence of tetrathionate gives it an advantage to successfully colonize the inflamed gut [69].

Although inflammation from the host allows *Salmonella* to thrive, too much inflammation can result in tissue damage which can trigger a severe immune response. To overcome this, *Salmonella* keeps host inflammation in check through GogB. Host ubiquitiniation systems regulate the host inflammatory response [70]. GogB is an effector protein in *Salmonella* that targets the host SCF E3 type ubiquitin ligase via an interaction with Skp1 and the human F-box only 22 (FBXO22) protein. SCF complexes of E3 ubiquitin ligase catalyze the ubiquination of proteins for degradation. The E3-type SCF ligase complex regulates the NFKB (nuclear transcription factor KB) pathway [71]. It has been shown that *Salmonella* lacking a functional GogB are unable to limit NFKB activation which causes an increase in the proinflammatory response in mice. This proinflammatory response is accompanied by extensive tissue damage and enhanced colonization of the gut in long-term chronic infection models [72]. GogB, then, is able to regulate the amount of inflammation-enhanced colonization by limiting the amount of tissue damage during infection [72].

METABOLIC GENES INVOLVED IN *SALMONELLA* PLANT COLONIZATION

In addition to being a problematic contaminant in meat and eggs, *Salmonella* contamination has also been an issue in produce where it has been linked to the presence of soft rot disease. Goudeau et al. showed that the population of *S.* Typhimurium increases 56-fold when inoculated by itself on to cilantro leaves, compared to a 2,884-fold increase in population size when co-inoculated with *Dickeya dadantii*, a pathogen that causes soft rot lesions in plants by macerating the plant tissue [73]. Transcriptome analysis of *D. dadantii* and *S.* Typhimurium indicate that *S.* Typhimurium uses anaerobic metabolism while infecting lettuce and cilantro and that it catabolizes substrates that are made available by the plant tissue degradation. Twenty-nine percent of the genes that were up-regulated in *S.* Typhimurium upon plant tissue infection had previously been shown to be involved in the colonization of the chick intestine [73]. Many of the genes identified in these studies were also shown to be involved in the colonization of mice, pigs, and cattle. The transcriptome data revealed that some of the most highly expressed genes in *S.* Typhimurium were genes involved in the utilization of 1,2-propanediol (*pdu* operon) and ethanolamine (*eut* operon) in both lettuce and cilantro [73]. Utilization of 1,2-propanediol and ethanolamine are dependent on cobalamin, and expression of genes involved in cobalamin synthesis are also highly expressed in *S.* Typhimurium upon colonization of produce. A *S.* Typhimurium mutant which was unable to utilize 1,2-propanediol or ethanolamine did not colonize lettuce and cilantro as efficiently as wild type indicating that these compounds represent significant carbon and energy sources for the bacterium when it colonizes produce [73].

It has also been shown that siderophore biosynthesis is required for efficient colonization of alfalfa by *Salmonella* [74]. Through screening deletion mutants, Hao et al. found that an *aroA* mutant was defective in alfalfa root colonization [74]. AroA is part of the chorismic acid biosynthesis pathway which is a metabolic part of aromatic amino acid and siderophore production. This growth defect was not complemented with the addition of tryptophan or phenylalanine, but growth and alfalfa colonization were restored with the addition of ferrous sulfate indicating that siderophore biosynthesis is important in the colonization of alfalfa roots. In support of this hypothesis, loss of *entB*, which is required for siderophore biosynthesis, interfered with the ability of *S.* Typhimurium to colonize alfalfa [74].

FUTURE DIRECTIONS

Salmonella continues to be a major foodborne pathogen and is transmitted primarily to humans through the consumption of meat, eggs or poultry. Infection of poultry and other food-producing animals with most *Salmonella* serovars generally results in asymptomatic intestinal colonization that leads to carcass contamination and entry into the food supply. If we are to begin to make progress on reducing the prevalence of *Salmonella* in food-producing animals and thereby reduce the incidence of salmonellosis in humans, we need to develop a better understanding of the metabolic processes in *Salmonella* which promote intestinal colonization of food animals by the pathogen. Results from recent –omics driven approaches have begun to close this critical gap in our knowledge by identifying a number of metabolic genes as having possible roles in *Salmonella* colonization of the chick, pig and calf. While roles for these metabolic genes in colonization will need to be verified, it seems likely that a significant number of these genes will prove to be important for efficient intestinal colonization of food-producing animals. Studies on how specific metabolic processes in *Salmonella* impact intestinal colonization have already yielded exciting discoveries, such as how *Salmonella* exploits the host's inflammation response to gain access to electron acceptors (e.g., nitrate and tetrathionate) which it can use for anaerobic respiration [65, 68]. We anticipate that future studies to characterize metabolic genes and their roles in colonization will similarly lead to exciting new discoveries regarding *Salmonella* physiology within the intestinal tract.

REFERENCES

[1] Brenner, F. W., Villar, R. G., Angulo, F. J., Tauxe, R. & Swaminathan, B. (2000). *Salmonella* nomenclature. *J Clin Microbiol*, *38*, 2465-2467.

[2] Scallan, E., Hoekstra, R. M., Angulo, F. J., Tauxe, R. V., Widdowson, M. A., Roy, S. L., Jones, J. L. & Griffin, P. M. Foodborne illness acquired in the United States--major pathogens. *Emerg Infect Dis*, *17*, 7-15.

[3] Frenzen, P., Riggs, T., Buzby, J., Breuer, T., Roberts, T., Voetsch, D., Reddy, S. & Group, F. W. (1999). *Salmonella* cost estimate update using FoodNet data. *Food Rev*, *22*, 10-15.

[4] Tauxe, R. V. (1991). *Salmonella*: a postmortem pathogen. *J Food Prot*, *54*, 563-568.

[5] Humphrey, T. J. (1994). Contamination of egg shell and contents with *Salmonella enteritidis*: a review. *Int J Food Microbiol*, *21*, 31-40.

[6] Althouse, C., Patterson, S., Fedorka-Cray, P. & Isaacson, R. E. (2003). Type 1 fimbriae of *Salmonella enterica* serovar Typhimurium bind to enterocytes and contribute to colonization of swine in vivo. *Infect Immun*, *71*, 6446-6452.

[7] Blondel, C. J., Yang, H. J., Castro, B., Chiang, S., Toro, C. S., Zaldivar, M., Contreras, I., Andrews-Polymenis, H. L. & Santiviago, C. A. (2010). Contribution of the type VI secretion system encoded in SPI-19 to chicken colonization by *Salmonella enterica* serotypes Gallinarum and Enteritidis. *PloS One*, *5*, e11724.

[8] Carnell, S. C., Bowen, A., Morgan, E., Maskell, D. J., Wallis, T. S. & Stevens, M. P. (2007). Role in virulence and protective efficacy in pigs of *Salmonella enterica* serovar Typhimurium secreted components identified by signature-tagged mutagenesis. *Microbiology*, *153*, 1940-1952.

[9] Clayton, D. J., Bowen, A. J., Hulme, S. D., Buckley, A. M., Deacon, V. L., Thomson, N. R., Barrow, P. A., Morgan, E., Jones, M. A., Watson, M. & Stevens, M. P. (2008). Analysis of the role of 13 major fimbrial subunits in colonisation of the chicken intestines by *Salmonella enterica* serovar Enteritidis reveals a role for a novel locus. *BMC Microbiol*, *8*, 228.

[10] Dhawi, A. A., Elazomi, A., Jones, M. A., Lovell, M. A., Li, H., Emes, R. D. & Barrow, P. A. (2011). Adaptation to the chicken intestine in *Salmonella* Enteritidis PT4 studied by transcriptional analysis. *Vet Microbiol*, *153*, 198-204.

[11] Dieye, Y., Ameiss, K., Mellata, M. & Curtiss R., 3rd. (2009). The *Salmonella* Pathogenicity Island (SPI) 1 contributes more than SPI2 to the colonization of the chicken by *Salmonella enterica* serovar Typhimurium. *BMC Microbiol*, *9*, 3.
[12] Harvey, P. C., Watson, M., Hulme, S., Jones, M. A., Lovell, M., Berchieri, A., Jr. Young, J., Bumstead, N. & Barrow, P. (2011). *Salmonella enterica* serovar Typhimurium colonizing the lumen of the chicken intestine grows slowly and upregulates a unique set of virulence and metabolism genes. *Infect Immun*, *79*, 4105-4121.
[13] Huang, Y., Leming, C. L., Suyemoto, M. & Altier, C. (2007). Genome-wide screen of *Salmonella* genes expressed during infection in pigs, using in vivo expression technology. *Appl Environ Microbiol*, *73*, 7522-7530.
[14] Morgan, E., Campbell, J. D., Rowe, S. C., Bispham, J., Stevens, M. P., Bowen, A. J., Barrow, P. A., Maskell, D. J. & Wallis, T. S. (2004). Identification of host-specific colonization factors of *Salmonella enterica* serovar Typhimurium. *Mol Microbiol*, *54*, 994-1010.
[15] Barrow, P. A., Hassan, J. O., Lovell, M. A. & Berchieri, A. (1990). Vaccination of chickens with *aroA* and other mutants of *Salmonella typhimurium* and *S. enteritidis*. *Res Microbiol*, *141*, 851-853.
[16] Porter, S. B., Tinge, S. A. & Curtiss, R., 3rd. (1993). Virulence of *Salmonella typhimurium* mutants for White Leghorn chicks. *Avian Dis*, *37*, 265-273.
[17] Rabsch, W., Methner, U., Voigt, W., Tschape, H., Reissbrodt, R. & Williams, P. H. (2003). Role of receptor proteins for enterobactin and 2,3-dihydroxybenzoylserine in virulence of *Salmonella enterica*. *Infect Immun*, *71*, 6953-6961.
[18] Zhang-Barber, L., Turner, A. K., Martin, G., Frankel, G., Dougan, G. & Barrow, P. A. (1997). Influence of genes encoding proton-translocating enzymes on suppression of *Salmonella typhimurium* growth and colonization. *J Bacteriol*, *179*, 7186-7190.
[19] Curtiss, R., 3rd & Kelly, S. M. (1987). *Salmonella typhimurium* deletion mutants lacking adenylate cyclase and cyclic AMP receptor protein are avirulent and immunogenic. *Infect Immun*, *55*, 3035-3043.
[20] Hensel, M., Shea, J. E., Gleeson, C., Jones, M. D., Dalton, E. & Holden, D. W. (1995). Simultaneous identification of bacterial virulence genes by negative selection. *Science*, *269*, 400-403.
[21] Hunt, T. A., Kooi, C., Sokol, P. A. & Valvano, M. A. (2004). Identification of *Burkholderia cenocepacia* genes required for bacterial survival in vivo. *Infect Immun*, *72*, 4010-4022.
[22] Karlyshev, A. V, Oyston, P. C., Williams, K., Clark, G. C., Titball, R. W., Winzeler, E. A. & Wren, B. W. (2001). Application of high-density array-based signature-tagged mutagenesis to discover novel *Yersinia* virulence-associated genes. *Infect Immun*, *69*, 7810-7819.
[23] Lehoux, D. E., Sanschagrin, F. & Levesque, R. C. (1999). Defined oligonucleotide tag pools and PCR screening in signature-tagged mutagenesis of essential genes from bacteria. *BioTechniques*, *26*, 473-478, 480.
[24] Bearson, S. M., Bearson, B. L. & Rasmussen, M. A. (2006). Identification of *Salmonella enterica* serovar Typhimurium genes important for survival in the swine gastric environment. *Appl Environ Microbiol*, *72*, 2829-2836.

[25] Bispham, J., Tripathi, B. N., Watson, P. R. & Wallis, T. S. (2001). *Salmonella* pathogenicity island 2 influences both systemic salmonellosis and *Salmonella*-induced enteritis in calves. *Infect Immun*, *69*, 367-377.

[26] Chan, K., Kim, C. C. & Falkow, S. (2005). Microarray-based detection of *Salmonella enterica* serovar Typhimurium transposon mutants that cannot survive in macrophages and mice. *Infect Immun*, *73*, 5438-5449.

[27] Geng, S., Jiao, X., Barrow, P., Pan, Z. & Chen, X. (2014). Virulence determinants of *Salmonella* Gallinarum biovar Pullorum identified by PCR signature-tagged mutagenesis and the *spiC* mutant as a candidate live attenuated vaccine. *Vet Microbiol*, *168*, 388-394.

[28] Hisert, K. B., MacCoss, M., Shiloh, M. U., Darwin, K. H., Singh, S., Jones, R. A., Ehrt, S., Zhang, Z., Gaffney, B. L., Gandotra, S., Holden, D. W., Murray, D. & Nathan, C. (2005). A glutamate-alanine-leucine (EAL) domain protein of *Salmonella* controls bacterial survival in mice, antioxidant defence and killing of macrophages: role of cyclic diGMP. *Mol Microbiol*, *56*, 1234-1245.

[29] Ku, Y. W., McDonough, S. P., Palaniappan, R. U., Chang, C. F. & Chang, Y. F. (2005). Novel attenuated *Salmonella enterica* serovar Choleraesuis strains as live vaccine candidates generated by signature-tagged mutagenesis. *Infect Immun*, *73*, 8194-8203.

[30] Lichtensteiger, C. A. & Vimr, E. R. (2003). Systemic and enteric colonization of pigs by a *hilA* signature-tagged mutant of *Salmonella choleraesuis*. *Microb Path*, *34*, 149-154.

[31] Pullinger, G. D., Paulin, S. M., Charleston, B., Watson, P. R., Bowen, A. J., Dziva, F., Morgan, E., Villarreal-Ramos, B., Wallis, T. S. & Stevens, M. P. (2007). Systemic translocation of *Salmonella enterica* serovar Dublin in cattle occurs predominantly via efferent lymphatics in a cell-free niche and requires type III secretion system 1 (T3SS-1) but not T3SS-2. *Infect Immun*, *75*, 5191-5199.

[32] Shah, D.H., Lee, M. J., Park, J. H., Lee, J. H., Eo, S. K., Kwon, J. T. & Chae, J. S. (2005). Identification of *Salmonella gallinarum* virulence genes in a chicken infection model using PCR-based signature-tagged mutagenesis. *Microbiology*, *151*, 3957-3968.

[33] Shea, J. E., Hensel, M., Gleeson, C. & Holden, D. W. (1996). Identification of a virulence locus encoding a second type III secretion system in *Salmonella typhimurium*. *Proc Natl Acad Sci USA*, *93*, 2593-2597.

[34] Thomsen, L. E., Chadfield, M. S., Bispham, J., Wallis, T. S., Olsen, J. E. & Ingmer, H. (2003). Reduced amounts of LPS affect both stress tolerance and virulence of *Salmonella enterica* serovar Dublin. *FEMS Microbiol Lett*, *228*, 225-231.

[35] Tsolis, R. M., Townsend, S. M., Miao, E. A., Miller, S. I., Ficht, T. A., Adams, L. G. & Baumler, A. J. (1999). Identification of a putative *Salmonella enterica* serotype typhimurium host range factor with homology to IpaH and YopM by signature-tagged mutagenesis. *Infect Immun*, *67*, 6385-6393.

[36] Seim, H., Loster, H., Claus, R., Kleber, H. P. & Strack, E. (1982). Stimulation of the anaerobic growth of *Salmonella typhimurium* by reduction of L-carnitine, carnitine derivatives and structure-related trimethylammonium compounds. *Arch Microbiol*, *132*, 91-95.

[37] Chaudhuri, R. R., Morgan, E., Peters, S. E., Pleasance, S. J., Hudson, D. L., Davies, H. M., Wang, J., van Diemen, P. M., Buckley, A. M., Bowen, A. J., Pullinger, G. D., Turner, D. J., Langridge, G. C., Turner, A. K., Parkhill, J., Charles, I. G., Maskell, D. J.

& Stevens, M. P. (2013). Comprehensive assignment of roles for *Salmonella typhimurium* genes in intestinal colonization of food-producing animals. *PLoS Gene*, *9*, e1003456.

[38] Wiame, E., Lamosa, P., Santos, H. & Van Schaftingen, E. (2005). Identification of glucoselysine-6-phosphate deglycase, an enzyme involved in the metabolism of the fructation product glucoselysine. *Biochem J*, *392*, 263-269.

[39] Baynes, J. W., Watkins, N. G., Fisher, C. I., Hull, C. J., Patrick, J. S., Ahmed, M. U., Dunn, J. A. & Thorpe, S. R. (1989). The Amadori product on protein: structure and reactions. *Prog Clin Biol Res*, *304*, 43-67.

[40] Suarez, G., Rajaram, R., Oronsky, A. L. & Gawinowicz, M. A. (1989). Nonenzymatic glycation of bovine serum albumin by fructose (fructation). Comparison with the Maillard reaction initiated by glucose. *J Biol Chem*, *264*, 3674-3679.

[41] Datsenko, K. A. & Wanner, B. L. (2000). One-step inactivation of chromosomal genes in *Escherichia coli* K-12 using PCR products. *Proc Natl Acad Sci USA*, *97*, 6640-6645.

[42] Ferenci, T. & Kornberg, H. L. (1973). The utilization of fructose by *Escherichia coli*. Properties of a mutant defective in fructose 1-phosphate kinase activity. *Biochem J*, *132*, 341-347.

[43] Kadner, R. J., Murphy, G. P. & Stephens, C. M. (1992). Two mechanisms for growth inhibition by elevated transport of sugar phosphates in *Escherichia coli*. *J Gen Microbiol*, *138*, 2007-2014.

[44] Yarmolinsky, M. B., Wiesmeyer, H., Kalckar, H. M. & Jordan, E. (1959). Hereditary defects in galactose metabolism in *Escherichia coli* mutants, II. Galactose-induced sensitivity. *Proc Natl Acad Sci USA*, *45*, 1786-1791.

[45] Kroger, C., Colgan, A., Srikumar, S., Handler, K., Sivasankaran, S. K., Hammarlof, D. L., Canals, R., Grissom, J. E., Conway, T., Hokamp, K. & Hinton, J. C. (2013). An infection-relevant transcriptomic compendium for *Salmonella enterica* serovar Typhimurium. *Cell Host Microbe*, *14*, 683-695.

[46] Steeb, B., Claudi, B., Burton, N. A., Tienz, P., Schmidt, A., Farhan, H., Maze, A. & Bumann, D. (2013). Parallel exploitation of diverse host nutrients enhances *Salmonella* virulence. *PLoS Path*, *9*, e1003301.

[47] Andrews, N. C. & Schmidt, P. J. (2007). Iron homeostasis. *Ann Rev Physiol*, *69*, 69-85.

[48] Raymond, K. N., Dertz, E. A. & Kim, S. S. (2003). Enterobactin: an archetype for microbial iron transport. *Proc Natl Acad Sci USA*, *100*, 3584-3588.

[49] Flo, T. H., Smith, K. D., Sato, S., Rodriguez, D. J., Holmes, M. A., Strong, R. K., Akira, S. & Aderem, A. (2004). Lipocalin 2 mediates an innate immune response to bacterial infection by sequestrating iron. *Nature*, *432*, 917-921.

[50] Berger, T., Togawa, A., Duncan, G. S., Elia, A. J., You-Ten, A., Wakeham, A., Fong, H. E., Cheung, C. C. & Mak, T. W. (2006). Lipocalin 2-deficient mice exhibit increased sensitivity to *Escherichia coli* infection but not to ischemia-reperfusion injury. *Proc Natl Acad Sci USA*, *103*,1834-1839.

[51] Muller, S. I., Valdebenito, M. & Hantke, K. (2009). Salmochelin, the long-overlooked catecholate siderophore of *Salmonella*. *Biometals*, *22*, 691-695.

[52] Fischbach, M. A., Lin, H., Liu, D. R. & Walsh, C. T. (2006). How pathogenic bacteria evade mammalian sabotage in the battle for iron. *Nat Chem Biol*, *2*, 132-138.

[53] Raffatellu, M., George, M. D., Akiyama, Y., Hornsby, M. J., Nuccio, S. P., Paixao, T. A., Butler, B. P., Chu, H., Santos, R. L., Berger, T., Mak, T. W., Tsolis, R. M., Bevins,

C. L., Solnick, J. V., Dandekar, S. & Baumler, A. J. (2009). Lipocalin-2 resistance confers an advantage to *Salmonella enterica* serotype Typhimurium for growth and survival in the inflamed intestine. *Cell Host Microbe*, *5*, 476-486.

[54] Henker, J., Laass, M., Blokhin, B. M., Bolbot, Y. K., Maydannik, V. G., Elze, M., Wolff, C. & Schulze, J. (2007). The probiotic *Escherichia coli* strain Nissle 1917 (EcN) stops acute diarrhoea in infants and toddlers. *Eur J Ped*, *166*, 311-318.

[55] Garcia, E. C., Brumbaugh, A. R. & Mobley, H. L. (2011). Redundancy and specificity of *Escherichia coli* iron acquisition systems during urinary tract infection. *Infect Immun*, *79*, 1225-1235.

[56] Deriu, E., Liu, J. Z., Pezeshki, M., Edwards, R. A., Ochoa, R. J., Contreras, H., Libby, S. J., Fang, F. C. & Raffatellu, M. (2013). Probiotic bacteria reduce Salmonella typhimurium intestinal colonization by competing for iron. *Cell Host Microbe*, *14*, 26-37.

[57] Corbin, B. D., Seeley, E. H., Raab, A., Feldmann, J., Miller, M. R., Torres, V. J., Anderson, K. L., Dattilo, B. M., Dunman, P. M., Gerads, R., Caprioli, R. M., Nacken, W., Chazin, W. J. & Skaar, E. P. (2008). Metal chelation and inhibition of bacterial growth in tissue abscesses. *Science*, *319*, 962-965.

[58] Liu, J. Z., Jellbauer, S., Poe, A. J., Ton, V., Pesciaroli, M., Kehl-Fie, T. E., Restrepo, N. A., Hosking, M. P., Edwards, R. A., Battistoni, A., Pasquali, P., Lane, T. E., Chazin, W. J., Vogl, T., Roth, J., Skaar, E. P. & Raffatellu, M. (2012). Zinc sequestration by the neutrophil protein calprotectin enhances *Salmonella* growth in the inflamed gut. *Cell Host Microbe*, *11*, 227-239.

[59] Flint, H. J., Bayer, E. A., Rincon, M. T., Lamed, R. & White, B. A. (2008). Polysaccharide utilization by gut bacteria: potential for new insights from genomic analysis. *Nature Rev Microbiol*, *6*, 121-131.

[60] Ng, K. M., Ferreyra, J. A., Higginbottom, S. K., Lynch, J. B., Kashyap, P. C., Gopinath, S., Naidu, N., Choudhury, B., Weimer, B. C., Monack, D. M. & Sonnenburg, J. L. (2013). Microbiota-liberated host sugars facilitate post-antibiotic expansion of enteric pathogens. *Nature*, *502*, 96-99.

[61] Spees, A. M., Wangdi, T., Lopez, C. A., Kingsbury, D. D., Xavier, M. N., Winter, S. E., Tsolis, R. M. & Baumler, A. J. (2013). Streptomycin-induced inflammation enhances *Escherichia coli* gut colonization through nitrate respiration. *mBio*, *4*.

[62] Rabsch, W., Tschape, H. & Baumler, A. J. (2001). Non-typhoidal salmonellosis: emerging problems. *Microbes Infect*, *3*, 237-247.

[63] Mirold, S., Rabsch, W., Rohde, M., Stender, S., Tschape, H., Russmann, H., Igwe, E. & Hardt, W. D. (1999). Isolation of a temperate bacteriophage encoding the type III effector protein SopE from an epidemic *Salmonella typhimurium* strain. *Proc Natl Acad Sci USA*, *96*, 9845-9850.

[64] Zhang, S., Santos, R. L., Tsolis, R. M., Mirold, S., Hardt, W. D., Adams, L. G. & Baumler, A. J. (2002). Phage mediated horizontal transfer of the *sopE1* gene increases enteropathogenicity of *Salmonella enterica* serotype Typhimurium for calves. *FEMS Microbiol Lett*, *217*, 243-247.

[65] Lopez, C. A., Winter, S. E., Rivera-Chavez, F., Xavier, M. N., Poon, V., Nuccio, S. P., Tsolis, R. M. & Baumler, A. J. (2012). Phage-mediated acquisition of a type III secreted effector protein boosts growth of *Salmonella* by nitrate respiration. *mBio*, *3*.

[66] Furne, J., Springfield, J., Koenig, T., DeMaster, E. & Levitt, M. D. (2001). Oxidation of hydrogen sulfide and methanethiol to thiosulfate by rat tissues: a specialized function of the colonic mucosa. *Biochem Pharm*, *62*, 255-259.

[67] Hinsley, A. P. & Berks, B. C. (2002). Specificity of respiratory pathways involved in the reduction of sulfur compounds by *Salmonella enterica*. *Microbiology*, *148*, 3631-3638.

[68] Winter, S. E., Thiennimitr, P., Winter, M. G., Butler, B. P., Huseby, D. L., Crawford, R. W., Russell, J. M., Bevins, C. L., Adams, L. G., Tsolis, R. M., Roth, J. R. & Baumler, A. J. (2010). Gut inflammation provides a respiratory electron acceptor for *Salmonella*. *Nature*, *467*, 426-429.

[69] Thiennimitr, P., Winter, S. E., Winter, M. G., Xavier, M. N., Tolstikov, V., Huseby, D. L., Sterzenbach, T., Tsolis, R. M., Roth, J. R. & Baumler, A. J. (2011). Intestinal inflammation allows *Salmonella* to use ethanolamine to compete with the microbiota. *Proc Natl Acad Sci USA*, *108*, 17480-17485.

[70] Cardozo, T. & Pagano, M. (2004). The SCF ubiquitin ligase: insights into a molecular machine. *Nature Rev Mol Cell Biol*, *5*, 739-751.

[71] Winston, J. T., Strack, P., Beer-Romero, P., Chu, C. Y., Elledge, S. J. & Harper, J. W. (1999). The $SCF^{\beta\text{-}TRCP}$-ubiquitin ligase complex associates specifically with phosphorylated destruction motifs in IκBα and β-catenin and stimulates IκBα ubiquitination in vitro. *Genes Dev*, *13*, 270-283.

[72] Pilar, A. V., Reid-Yu, S. A., Cooper, C. A., Mulder, D. T. & Coombes, B. K. (2012). GogB is an anti-inflammatory effector that limits tissue damage during *Salmonella* infection through interaction with human FBXO22 and Skp1. *PLoS Path*, *8*, e1002773.

[73] Goudeau, D. M., Parker, C. T., Zhou, Y., Sela, S., Kroupitski, Y. & Brandl, M. T. (2013). The *Salmonella* transcriptome in lettuce and cilantro soft rot reveals a niche overlap with the animal host intestine. *Appl Environ Microbiol*, *79*, 250-262.

[74] Hao, L. Y., Willis, D. K., Andrews-Polymenis, H., McClelland, M. & Barak, J. D. (2012). Requirement of siderophore biosynthesis for plant colonization by *Salmonella enterica*. *Appl Environ Microbiol*, *78*, 4561-4570.

In: Salmonella
Editor: Christopher B. Hackett

ISBN: 978-1-63463-651-3

Chapter 8

MOLECULAR TYPING SCHEMES OF FOOD-ASSOCIATED SALMONELLAE

Spiros Paramithiotis**, Agni Hadjilouka and Eleftherios H. Drosinos

Laboratory of Food Quality Control and Hygiene, Department of Food Science and Human Nutrition, Agricultural University of Athens, Athens, Greece

ABSTRACT

Salmonella is the confirmed causative agent of numerous foodborne outbreaks throughout the world. Successful monitoring of the outbreaks and effective implementation of control programs essentially requires the accurate identification of the infection source and the transmission pathways throughout the food chain. Traditional typing relied on phenotypic characteristics such as serotyping, phage typing and antimicrobial susceptibility testing. More recently a wide range of methods relying on differences at genome level, such as Pulsed-Field Gel Electrophoresis (PFGE), Multi-Locus Variable number of tandem repeats Analysis (MLVA), Multi Locus Sequence Typing (MLST), Single Nucleotide Polymorphisms (SNPs) of Whole Genome Sequences (WGS) and Clustered Regularly Interspaced Palindromic Repeats (CRISPR) have been introduced, thoroughly studied and extensively applied. All these techniques are characterized by certain advantages and drawbacks that should be taken into consideration before any conclusion of epidemiological nature is drawn. In this chapter all information relevant to current typing schemes of food-associated salmonellae is integrated and critically reviewed.

INTRODUCTION

Salmonella is the causative agent of hundreds of millions of human illnesses and hundreds of thousands of deaths annually worldwide. In the U.S. a number of approx. 1.0

* Corresponding author. E-mail: sdp@aua.gr; Tel.: +30 210 5294705 Fax: +30 210 5294683

million salmonellosis cases are reported annually; it is the second most frequent causative agent of foodborne illness with Norovirus being the most frequent. The estimated annual hospitalizations account for approx. 19.000 with approx. 380 deaths. Regarding the latter, salmonellosis is responsible for more deaths than any other foodborne illness, on an annual base. The non-typhoidal serotypes Typhimurium, Enteritidis, Newport, Heidelberg and Javiana were the 5 most frequently reported serotypes isolated from human sources from 1999 to 2009; the typhoidal serotypes Paratyphi B var. L(+) tartrate+ and Typhi were in the 20 most frequently reported serotypes isolated from human sources during the same decade (CDC, 2012). *Salmonella* in food was mainly associated with meat and meat product; however presence on fruits and vegetables is increasingly reported as well as the respective outbreaks.

A plethora of subtyping techniques currently exists. Epidemiological surveillance involves the identification of the source(s) of contamination as well as the route of transmission. In that context, subtyping is needed to assist, integrate and verify epidemiological data. In the absence of such data, great caution should be given to the interpretation of subtyping as it is very likely to lead to erroneous conclusions. A genetic marker is selected in order to provide with the above mentioned support. Therefore, the selection of the genetic marker is of paramount importance. First of all, it should be stable during the study period. This means that selection of different markers may be necessary for short-term and long-term studies. Moreover, the genetic marker should be able to separate the isolates and at the same time remain concordant with the epidemiological data. Regarding the technique selected to assess this genetic marker; it should lead to reproducible results irrespective operator, place and time and therefore facilitate data exchange (van Belkum et al., 2007).

The typing schemes that have been used for *Salmonella* include both phenotypic and genotypic approaches. Serotyping, phage-typing and antimicrobial resistance typing belong to the former. The latter may be further subdivided according to the type of the data that is compared to profile-based analysis and sequence-based analysis. Ribotyping, plasmid profile analysis, Pulsed-Field Gel Electrophoresis (PFGE) and Multi-Locus Variable-Number Tandem Repeat Analysis (MLVA) belong to the former while Multi-Locus Sequence Typing (MLST), Single Nucleotide Polymorphisms (SNPs) analysis, Whole Genome Sequence (WGS) analysis and Clustered Regularly Interspaced Short Palindrome Repeats (CRISPRs) analysis belong to the latter. Serotyping has been the most widely used subtyping method. It is used to study the antigenic variability in the outer membrane lipopolysaccharides (O antigens), flagellar proteins (H1 and H2 antigens) and capsular polysaccharide (Vi antigens). This scheme has created a reference discrimination for *Salmonella* to over 2500 serovars. Due to the technical limitations of traditional serotyping, PCR-based protocols have been proposed (Kim et al., 2006; Malorny et al., 2009; Leader et al., 2009; Liu et al., 2011a; Beaubrun et al., 2012) assessing the most common serotypes. From an epidemiological perspective, serotyping is of limited usefulness since the serotypes associated with human disease are very few. Phage typing is another typing scheme in which the strains are characterized by their reactivity against a set of phages. Although it is not technically demanding, it is prone to interpretation errors and not fully reproducible. However, it is useful for an initial characterization of isolates and of similar usefulness to the serotyping scheme regarding epidemiological surveillance. Antimicrobial resistance typing determines the resistance profile of a strain to a set of antimicrobials. It has also been extensively used in the past and is still employed in some studies. It is not technically demanding and of similar usefulness with the previous phenotypic methods.

Regarding the genotyping approaches that have been used for *Salmonella* subtyping, PFGE is the preferred one by the majority of the laboratories that study *Salmonella* from an epidemiological perspective. PFGE is currently recognized as one of the most effective typing techniques for clustering and differentiation generally of foodborne pathogens. In the case of *Salmonella*, numerous studies employing this technique in epidemiological surveillance have been published. However, typing with PFGE is not effective in all cases and assistance by other techniques is sometimes required and is currently available, as will be presented in the following paragraphs.

PROFILE-BASED COMPARISONS

In profile-based comparisons the isolates are subjected to a genotypic profile-generating procedure, either PCR mediated (e.g. MLVA) or not (e.g. PFGE). Differences in the sequences between the isolates are detected and visualized through the resulting genotypic profile. Two isolates are considered identical when the number and the apparent sizes of the fragments are identical. In the case of PFGE, isolates are classified as closely related or possibly related when the observed number of fragment differences are 2-3 and 4-6, respectively. On the other hand, when that number is more than seven, the isolates are characterized as unrelated (Tenover et al., 1995).

The most effective profile-based techniques, from an epidemiological perspective are PFGE and MLVA that are further discussed.

Pulsed-Field Gel Electrophoresis (PFGE)

The Pulsed-Field Gel Electrophoresis includes the use of restriction enzymes to cleave DNA into large fragments that are subsequently separated with gel electrophoresis in a pulsed field. It has been extensively used for the assessment of genetic diversity among various isolates from a wide range of isolation sources (Paramithiotis et al., 2008; Doulgeraki et al., 2010; 2011) as well as for subtyping of bacterial pathogens. Regarding the latter, it is currently considered as the 'gold standard' due to its high discriminating power and good epidemiological concordance. Therefore, an internet-based platform, namely PulseNet, has been created to enable public health laboratories to share PFGE data and thus identify outbreaks (Swaminathan et al., 2001).

The discriminatory power of PFGE may be adjusted by selecting the appropriate type and number of endonucleases. In cases in which application of an enzyme results in epidemiologically meaningless clusters or fails to separate unrelated isolates, then additional enzymes should be employed, e.g. PulseNet recommends primarily the use of *Asc*I for *L. monocytogenes* surveillance and secondarily the use of *Apa*I. On the contrary, cases in which PFGE is too discriminatory, i.e. fails to cluster epidemiologically related isolates, are not very often reported and may occur in long-term studies that include isolates that share a very recent common ancestor.

Regarding subtyping of *Salmonella* isolates, PulseNet recommends primarily the use of *Xba*I, secondarily of *Bln*I/*Avr*II and tertiary of *Spe*I. Many outbreaks have been identified by PFGE but there are also many cases in which assistance by another technique was necessary in order to separate indistinguishable by PFGE isolates; some of the proposed technique combinations are discussed further in the text.

PFGE is also characterized by certain disadvantages; it is slow, labor-intensive, requires appropriately trained personnel and the results are often prone to misinterpretation. Especially regarding the latter, much effort has been made to reach a consensus. The first effective attempt was made by Tenover et al. (1995); a glossary of terms as well as the categories of genetic and epidemiologic relatedness was set. Moreover, guidelines were given for the interpretation of the PFGE patterns within the context of epidemiological studies of outbreaks spanning relatively short periods. The latter has been proved very important since the interpretation in the context of organism diversity must also take into consideration the clonality of the particular microorganism under study and therefore may be totally different. An update to these recommendations for interpreting PFGE results was provided by Barrett et al. (2006) in an attempt to address the inconsistencies after years of application.

Multi-Locus Variable-Number Tandem Repeat Analysis (MLVA)

The Multi-Locus Variable-Number Tandem Repeat Analysis (MLVA) involves the comparison of multiple variable-number tandem-repeat sequences (VNTRs) of DNA between strains. Clustering capacity can be adjusted by varying the number and the type of the loci that are studied; rapidly evolving VNTRs are useful for the discrimination of closely related isolates and in short-term epidemiological studies whereas slowly evolving VNTRs may be used to assess genetic relatedness among isolates with different evolutionary distances and in long-term epidemiological assessment. MLVA generally exhibits both relative stability and concordance with PFGE. Moreover, it has important advantages: the protocol is less labor-intensive, faster and much cheaper than PFGE. Furthermore, it is characterized by higher processing capacity, it can be completely automated and the generated data are more easily analyzed and shared between laboratories. Therefore several MLVA protocols have been developed and successfully used for subtyping of many bacteria such as *Bacillus anthracis* (Keim et al., 2000), *Borderella pertussis* (Schouls et al., 2004), *Coxiella burnetti* (Svraka et al., 2006), *Enterococcus faecalis* (Titze-de-Almeida et al., 2004), *Escherichia coli* O157:H7 (Lindstedt et al., 2003b; Noller et al., 2003), *Francisella tularensis* (Farlow et al., 2001; Johansson et al., 2004), *Haemophilus influenzae* (Schouls et al., 2005), *Mycobacterium tuberculosis* (Allix et al., 2004), *M. paratuberculosis* (Overduin et al., 2004), *Neisseria meningitidis* (Yazdankhah et al., 2005), *Pseudomonas aeruginosa* (Outeniente et al., 2003), *Shigella sonnei* (Liang et al., 2007) and *Yersinia pestis* (Klevytska et al., 2001).

As far as *Salmonella* is concerned, MLVA is serotype specific; therefore each serotype requires the evaluation of new loci. For that purpose many schemes targeting at various VNTRs have been developed over the years.

Salmonella Typhimurium has been the epicenter of intensive study regarding the development of an effective MLVA protocol due to the significance of this serotype in foodborne salmonellosis as well as the inability of PFGE to discriminate between certain closely related strains, particularly DT104 isolates. Lindstedt et al. (2003a) developed a

VNTR-based typing assay using 78 isolates. The use of loci designated as STTR1 to STTR8 provided with better resolution than PFGE with *Xba*I. This protocol was further improved by Lindstedt et al. (2004). A better adaptation to capillary electrophoresis was performed in that study by excluding STTR1, STTR2, STTR4, STTR7 and STTR8, retaining STTR3, STTR5 and STTR6 and adding two new loci designated as STTR9 and STTR10 with more appropriate size ranges for capillary electrophoresis. The improved protocol was effectively applied in the subtyping of 106 *S.* Typhimurium isolates of various phage types. Furthermore, it was successfully applied by Hopkins et al. (2007) to analyze 190 isolates from 8 epidemiologically distinct outbreaks. It was concluded that MLVA provided with better discrimination than PFGE with *Xba*I for some phage types. More accurately, isolates from three outbreaks belonging to phage types 104 and 104b that were indistinguishable by PFGE were successfully allocated to the outbreaks with MLVA. Moreover, the results suggested the possibility of small changes within certain loci during the course of an outbreak. Ross et al. (2009) assayed 78 *S.* Typhimurium isolates, from which 37 were DT126 and 41 non-DT126. MLVA was performed on 10 loci; STTR2, STTR3, STTR5, STTR6, STTR7, STTR9 and STTR10 according to the studies by Lindstedt et al. (2003a; 2004) and TR1, TR2 and TR3 according to the study by Liu et al. (2003) on *S.* Typhi. Generally, loci STTR5 and STTR6 were reported as the most variable. However, the variability of locus STTR5 was only observed for the 41 non-DT126 isolates. On the contrary, no variation in fragment length was observed regarding DT126 isolates. These differences were assigned to variation in the genetic make-up of different phage types regarding the evolution of the loci used for the MLVA as well as to the origin of the isolates. As far as the loci that were derived from the studies on *S.* Typhi were concerned, they failed to provide any discrimination between the *S.* Typhimurium isolates. A total of 440 *S.* Typhimurium isolates collected from a variety of sources over a period of 15 years were used for the development and evaluation of a MLVA protocol by Chiou et al. (2010). A number of 16 VNTRs, designated as ST01 (STTR7), ST06 (STTR3), ST16, ST17, ST19 (STTR6), ST20, ST22, ST23, ST25 (STTR5), ST26 (STTR9), ST28, ST30, ST35, ST36, ST38 and ST40 (STTR10) were tested for their ability to cluster and differentiate the isolates. STTR5, STTR10, STTR6 and STTR3 were reported to be the most diverse loci while ST35, ST36, ST38 and ST30 were the less diverse. MLVA based on the 4 most variable VNTRs (MLVA4) exhibited better resolving power than PFGE with *Xba*I whereas MLVA8 resulted in additional discriminatory power. The clonal relationships between the isolates were better displayed by the MLVA based on all 16 VNTRs.

Salmonella Enteritidis is reported as the second most common *Salmonella* serotype in the U.S. Moreover, PFGE has only limited usefulness in epidemiological surveillance of this serotype due to the fact that this is one of the most genetically homogeneous serotypes. Therefore, the need for the development of an MLVA protocol was imperative. Boxrud et al. (2007) selected 10 loci, designated as SE1 to SE10 and assayed one hundred fifty-three isolates, forty of which from four separate outbreaks and the rest without any known epidemiological link. MLVA provided with better discrimination than PFGE with *Xba*I and *Bln*I, as indicated by the Simpson's index. Regarding epidemiological concordance, both MLVA and PFGE correctly clustered the isolates within the respective outbreaks. However, MLVA allowed better discrimination of isolates between outbreaks as well as between outbreak-related and unrelated ones. A multiplex PCR targeting seven VNTRs, namely SE1, SE2, SE3, SE5, SE7, SE8 and SE9 coupled with capillary electrophoresis was developed by Cho et al. (2007). A total of 34 *S.* Enteritidis isolates from various sources was assayed. It was reported that MLVA had better discriminatory power than PFGE with *Xba*I and *Bln*I. Moreover, the clonal groups based on MLVA were largely associated with the origin of the isolates. Cho et al. (2008) studied the allele distribution and the genetic diversity of nine VNTR loci between 145 isolates from different sources including humans, chickens and eggs. The loci described by Boxrud et al. (2007) were selected except for SE4. It was revealed that the genetic diversity, as defined by Nei's index, was higher in human isolates (0.41) than chicken (0.30) and egg ones (0.16). Especially regarding loci SE3, SE7 and SE9, human isolates displayed significantly higher diversity than the isolates from the other sources. On the contrary, chicken isolates exhibited significantly higher diversity in loci SE5 and SE10 than the isolates from the other sources. A new MLVA method with emphasis on the most predominant phage types PT4 and PT8 was reported by Malorny et al. (2008). A set of nine loci was amplified by multiplex PCR; fragment size analysis was performed automatically using a multicolor capillary electrophoresis instrument. The loci selected were SE1, SE2, SE3, SE5, SE7, SE9 and SE10 according to Boxrud et al. (2007), STTR7 according to Lindstedt et al. (2003a) and the newly selected SENTR3. A total of 240 isolates including 23 definite phage types were assayed and excellent epidemiological concordance was exhibited. A simplified subtyping protocol based on three hypervariable VNTR loci was reported by Beranek et al. (2009). The loci selected were SE1 and SE2 according to Boxrud et al. (2007), STTR5 according to Lindstedt et al. (2003a) and ENTR6. One hundred twelve random isolates were included in the study. The MLVA provided with improved discrimination power than PFGE with *Xba*I, the former resulted in a Simpson's index of 0.91 while the latter in only 0.41.

S. Typhi and Paratyphi A are highly monomorphic and genetically related serovars (McClelland et al., 2004). Therefore PFGE may not be discriminatory enough to differentiate between outbreak related and unrelated strains and thus the development of an MLVA protocol was necessary. The development of an MLVA protocol for typing of *S.* Paratyphi A was reported by Tien et al. (2011). A total of 55 isolates were included in the study, collected in 10 years from at least 6 countries of South and Southeast Asia. The MLVA protocol was based on the study of 6 loci, namely Sal02 (Ramisse et al., 2004) and STTR5 (Lindstedt et al., 2004) and Sty14, Sty16, Sty19 and Sty26, and revealed the high clonality of the isolates. Moreover, it provided with higher discrimination level that PFGE with *Xba*I and *Bln*I. However, the level of discrimination was not sufficient enough to separate all epidemiologically unrelated isolates; situation was improved when the two methods were combined. Regarding *S.* Typhi, Liu et al. (2003) published a protocol based on a multiplex PCR of three VNTRs designated as TR1, TR2 and TR3. A total of 59 isolates were assayed revealing a high level of profile heterogeneity. Octavia and Lan (2009) reported the development of an MLVA assay employing nine VNTRs and coupled with capillary electrophoresis. The loci studied were the newly selected TR4500 and TR4699, TR1 and TR2 according to Liu et al. (2003) and the SAL02, SAL06, SAL10, SAL16 and SAL20 according to Remisse et al. (2004). A total of 73 isolates were assayed and a nearly complete discrimination was achieved. Although typing of the five most variable VNTRs, namely TR2, TR4699, SAL02, TR1 and SAL20 was sufficient to achieve maximum differentiation of the isolates included in the study, the addition of TR4500 and SAL16 was suggested to gain extra discriminatory power for outbreak investigations. The evolutionary relationships revealed by MLVA were in conflict with the ones obtained using SNP analysis (Octavia and Lan, 2007) most probably due to the increased speed with which VNTRs seem to have evolved and therefore the insufficient phylogenetic information that has been retained for the accurate depiction of the evolutionary relationships suggesting that VNTRs could be used to determine the relationships of only closely related isolates. The subtyping of 125 *S.* Typhi isolates collected in 14 years by MLST using 11 VNTRs, namely the newly selected Sty2 and Sty3, TR1, TR2 and TR5 (Liu et al., 2003), STTR5 (Lindstedt et al., 2003a), Sal02, Sal06 and Sal20 (Ramisse et al., 2004) and TR4500 and TR4699 (Octavia & Lan, 2009) was reported by Tien et al. (2012). Seven of the eleven VNTPs were highly variable and therefore MLVA8 displayed a high level of discrimination between the isolates without any epidemiological connection; the discrimination index was even higher than the respective obtained by PFGE with *Xba*I and *Bln*I. Moreover, establishment of accurate genetic relationships between very closely related *S.* Typhi isolates could be performed by using the whole set of VNTRs.

The development and application of an MLVA protocol for *S.* Newport, a clinically and epidemiologically important serovar that is also characterized by multidrug resistance, was reported by Davis et al. (2009). A set of 132 isolates was assayed by six VNTPs, four of which were specifically selected for *S.* Newport and two, namely STTR5 and STTR6 were derived from *S.* Typhimurium experiments (Lindstedt et al., 2003a). It was reported that the protocol demonstrated excellent repeatability and close congruence with PFGE with *Xba*I as well as strong correspondence with epidemiological data.

Bergamini et al. (2011) reported the development of an MLVA protocol for *S.* Gallinarum, the causative agent of fowl typhoid, a severe systemic disease of chicken that may cause heavy economic losses to the poultry industry. Six loci were used, four of which were specifically selected for *S.* Gallinarum, locus STTR10pl was derived from studies on *S.* Typhimurium (Lindstedt et al., 2004) and locus SE1 that was reported for *S.* Enteritidis (Cho et al. 2007). A total of 50 isolates were assayed and compared with the discrimination obtained by PFGE using *Xba*I, *Spe*I and *Not*I. The discrimination index obtained by MLVA was higher than the respective of PFGE that occurred from the combination of the three enzymes. Interestingly, *Xba*I provided with less discrimination than the other enzymes.

A combination of MLVA with multiple amplification of prophage locus typing (MAPLT) was proposed by Young et al. (2012) for differentiation and typing of closely related *S.* Heidelberg isolates. A set of 7 loci were selected for MAPLT and three for MLVA, namely STTR3, STTR5 and Sal20. Application of PFGE with *Xba*I on 64 isolates resulted in very poor differentiation, especially of the predominant PT1 phenotype. On the contrary, PT1 isolates were effectively separated by the proposed combination. Lettini et al. (2014) reported on the genetic relationship between *S.* Typhimurium and *S.* 4,[5],12,i:− belonging to the DT7a. A total of 74 isolates were analyzed, 24 of which were of human, food and animal origin and were collected during an outbreak in 2011while the rest were unrelated to the outbreak. MLVA was performed according to Lindstedt et al. (2004). PFGE with *Xba*I was not capable to distinguish 80% of the isolates. On the contrary, MLVA provided with better discrimination power separating nearly all outbreak related isolates from unrelated ones.

SEQUENCE-BASED COMPARISONS

The rationale behind the sequence-based comparisons is not much different from the respective of profile-based comparisons. Generally, the differences in the genome between isolates are assessed directly through sequencing of either specific loci or the whole genome. However, in some cases, such as the detection of some SNPs, there is hardly a difference in the methodological approach between the two types of comparison.

Sequencing-based comparisons have the potential to become the ultimate approach for epidemiological studies due to the apparent discriminatory power and portability of the results. However, there are many drawbacks that limit the use of this approach. The most important is the need for detection of polymorphic sequences with the appropriate discriminatory power. This task is still very expensive, labor-intensive and slow.

Multi-Locus Sequence Typing (MLST)

Multi-locus sequence typing (MLST) is an improved version of a previously commonly applied method called Multi-Locus Enzyme Electrophoresis (MLEE). In the latter, differences between isolates were assessed on the basis of enzyme polymorphisms detected through different electrophoretic mobilities. In MLST these differences are assessed directly in the nucleotide sequence. Selection of proper number and type of genes as markers allows the adjustment of the discriminatory power.

MLST schemes have been created for many organisms, including *Bartonella henselae* (Iredell et al., 2003), *Bordetella pertussis* (van Loo et al., 2002), *Campylobacter jejuni* (McTavish et al., 2009), *Edwardsiella tarda* (Yang et al., 2013), *Listeria monocytogenes* (Revazishvili et al., 2004; Parisi et al., 2010), *Mycobacterium abscessus* (Jeon et al., 2014), *Neisserria meningitidis* (Maiden et al., 1988), *Staphylococcus aureus* (Feil et al., 2003), *Vibrio cholerae* (Kotetishvili et al., 2003) and *Streptococcus suis* (King et al., 2002).

Regarding *Salmonella*, evaluation of some schemes has taken place and the discriminatory power is close to the respective of classical serotyping, which in no case is adequate for outbreak analysis. However, it may accurately depict the genetic relatedness in long-term epidemiological studies.

A total of 243 *Salmonella* isolates were characterized by MLST of 16S rRNA gene, *manB*, *glnA* and *pduF* by Kotetishvili et al. (2002). The most polymorphic gene was *manB* (5.6%) and the less was 16S rRNA (0.6%). It was reported that MLST provided with better discriminatory ability than PFGE and successfully differentiated isolates that were grouped in a single PFGE cluster. Topdahl et al. (2005) compared the effectiveness of MLST, PFGE and AFLP in subtyping of 110 *S. enterica* isolates of 25 serotypes from human and veterinary sources. MLST was based on *aroC*, *dnaN*, *hemD*, *hisD*, *purE*, *sucA* and *thrA*, PFGE was performed with *Xba*I and AFLP with *Bgl*II and *Bsp*DI. In general, all methods were successful in differentiating the isolates between the serotypes and to a lesser extend within the serotypes. However, MLST resulted in a lower discriminatory power compared to both AFLP and PFGE due to the low level of diversity exhibited by the genes, i.e. each gene contained between 3.6% and 6.4% polymorphic sites. By comparing this study with the one by Kotetishvili et al. (2002), although similar level of polymorphism was reported for the genes selected, the final outcome, in terms of discrimination was rather opposite, most probably due to differences in the PFGE protocol and the set of isolates included in the two studies.

An alternative MLST scheme was developed by Sukhnanand et al. (2005). The target genes were *panB*, *fimA*, *icdA*, *manB*, *spaN*, *mdh* and *aceK* and the scheme was evaluated on

66 isolates belonging to 12 serotypes. The highest subtype discrimination was obtained by *manB*, *fimA* and *mdh*; this set of genes separated the 66 isolates into 25 sequence types that correlated well with serotypes and additionally allowed the differentiation within the 9 of the 12 serotypes. On the contrary, Fakhr et al. (2005) highlighted the uselessness of *manB*, *pduF*, *glnA*, and *spaM* for the discrimination of *S.* Typhimurium clinical isolates. A total of 85 isolates were included in this study; the isolates were affectively separated by PFGE with *Xba*I into fifty clusters. On the contrary, sequencing of these four genes revealed no genetic diversity, i.e. 100% identity, and therefore no discrimination power was obtained.

The gene set used by Topdahl et al. (2005) has prevailed among the laboratories that are using MLST. The diversity of 121 isolates collected from food, feed and live chicken samples, from 13 regions or countries, was evaluated by Liu et al. (2011d). The highest level of diversity was revealed for *hisD*. Based on the Simpson's index, the discriminatory power of MLST was close to the one of serotyping, which exhibited an association between the two techniques but on the other hand failed to associate the isolates with the source of isolation or antimicrobial resistance. Almeida et al. (2013) studied a total of 35 *S.* Infantis isolates collected from human feces and food items from various cities in the Sao Paulo State, using the MLST scheme as well as ERIC-PCR and PFGE. The high clonality among the isolates was best exhibited by MLST and all isolates presented the same ST32. Application of ERIC-PCR and PFGE revealed a discrimination index of 0.71 and 0.96, respectively. Similar results were reported during the molecular epidemiology assessment of 95 *S.* Kottbus isolates collected from humans, food and animals in Germany (Toboldt et al., 2014). MLST revealed the high clonality among the isolates by identifying four sequence types. PFGE with *Xba*I and *Bln*I separated the isolates into 15 and 22 clusters, respectively. Conclusions of epidemiological nature, such as the possible sources of human infection could only be drawn by combining these results. Forty-two strains of *S.* Livingstone mostly of human origin (31 isolates from Tunisia and 9 from Belgium) were subtyped by MLST (Guedda et al., 2014). The majority of the isolates exhibited identical allelic profile and belonged to the ST543 sequence type. This could be attributed to the persistence of this clone to the given environment. Indeed, previous studies by Ktari et al. (2006; 2009) revealed that a single clone of *S.* Livingstone was responsible for inter-hospital outbreaks in different hospitals in Tunisia and therefore this sequence type may reflect this persistence. Strains from different origins (i.e. human origin and food products) were effectively separated by MLST in this study. Harbottle et al. (2006) compared MLST and PFGE typing for the characterization of 81 *S.* Newport isolates collected from various sources. PFGE with *Xba*I generated 43 patterns, whereas MLST resulted in 12 sequence types, one of which contained the 62% of the isolates under study. Although there was a general agreement between the two techniques, i.e. isolates clustered in closely related PFGE patterns were clustered in the major MLST sequence type while distantly related isolates were also separated by PFGE, accurate differentiation between them was only achieved by combining these two techniques. This moderate differentiation capacity of MLST was attributed to the selection of the housekeeping genes used and the probable slow accumulation of mutations within them. The phylogenetic utility of the MLST approach to discriminate between members of a closely related collection of salmonellae, namely the *Salmonella* reference collection A (SARA) that consists of 72 strains, referred to as the Typhimurium complex, belonging to *S.* Typhimurium and its four closest serological relatives, was assessed by Bell et al. (2011). It was concluded that current MLST scheme was capable of partitioning this group. More accurately, phylogenetic analysis revealed seven distinct evolutionary clades that assorted the strains largely along the serological lines. However, study of additional gene sets seem to be necessary in order to reveal more details regarding their evolution.

An alternative to classic MLST was proposed by Tankouo-Sandjong et al. (2007). The proposed scheme consisted of two housekeeping genes, *gyrB* and *atpD*, in combination with the flagellin genes *fliC* and *fljB*. This approach was termed MLST-v, for multilocus sequence typing based on virulence genes, to distinguish it from MLST based solely on housekeeping genes. This scheme was evaluated on multiple isolates belonging to 22 serotypes. The results obtained indicated that this approach allowed efficient molecular subtyping for at least the more common serotypes. Inclusion of *fliC* gene in the MLST scheme was of particular importance since it allowed detection of the emergence of new serotypes and an accurate assessment of the genetic relationship with their immediate predecessor that is not recognized by classical serotyping.

An advancement of MLST, termed MLST-seq was presented by Singh et al. (2012). The increased sequencing capacity of NGS was utilized in order to increase the target genes of MLST and therefore increase the discrimination power. For that purpose, a hairpin-primed multiplex amplification protocol was developed and used to analyze 20 target loci including housekeeping genes and virulence genes from 41 *S. enterica* isolates belonging to 9 serotypes. Analysis of the sequence data revealed that in each strain the consensus target sequence was missing for 2–14 target genes. More accurately, for the genes *thrA*, *fliC* and *icdA*, no assembled target sequence was obtained for the majority of the strains, whereas the only target genes for which the consensus sequences were obtained for all 41 strains were *sucA* and *atpD*. Despite that, advances in NGS may allow MLST-seq analysis to become a more practical and economically feasible way for strain subtyping.

A novel MLST approach for subtyping of the nine major *Salmonella* serovars employing two virulence genes, namely *fimH* and *sseL* and 2 CRISPR loci was proposed by Liu et al. (2011b; 2011c). Addition of CRISPRs significantly increased discriminatory power compared to previously published MLST schemes, especially regarding the highly clonal *S.* Enteritidis. The proposed MLST scheme exhibited high epidemiologic concordance for subtyping the major serovars of *Salmonella*, except for *S.* Muenchen, most probably due to the increased evolving speed of the CRISPRs of *S.* Muenchen that renders them inapplicable for outbreak investigation. The same approach was effectively used for subtyping of *S.* Typhimurium and *S.* Heidelberg (Shariat et al., 2013a), *S.* Newport (Shariat et al., 2013b) as well as *S.* Enteritidis (Shariat et al., 2013c) and its applicability and complementarity to PFGE has been demonstrated.

The replacement of classical phenotypic serotyping by MLST was argued by Achtman et al. (2012). The driving force to that replacement would be the level of information gained by both techniques and their possible utilization for epidemiological purposes. For that purpose, a total of 4257 isolates from 554 serovars were studied by MLST and effectively separated into 1092 sequence types. Between these sequence types a gradient of genetic relatedness has been recognized. Many of these closely related clusters corresponded to a particular serovar, such as most Typhimurium and most Enteritidis that were assigned to specific clusters, but many clusters contained more than one serovar. Furthermore, most serovars were polyphyletic and thus distributed across multiple genetically unrelated groups. Thus, classical serovar designations confounded genetically unrelated isolates and failed to recognize natural evolutionary groupings.

Single Nucleotide Polymorphisms (SNPs)

Single Nucleotide Polymorphisms are DNA sequences that differ at a single base between isolates. Identification of the SNPs that will offer the appropriate separation level, either in studies of epidemiological nature or in the assessment of evolutionary relationships, is a real challenge. The discrimination power may be adjusted by the selection of the appropriate type and number of SNPs.

SNP typing has been successfully applied for the characterization of *E. coli* O157:H7 (Manning et al., 2008; Zhang et al., 2006) and *L. monocytogenes* (Ducey et al., 2007; Ward et al., 2008). Regarding *Salmonella* it has been effectively applied to study the genetic relationships between *S.* Typhi isolates (Octavia and Lan, 2007). Moreover, based on these results, a SNP typing method using restriction enzyme digestion to detect 38 genome-wide SNPs as markers was developed that could potentially assist in epidemiological studies of this serovar. Furthermore, the same authors developed a hairpin primer real-time assay for SNP typing of *S.* Typhi isolates. A set of 42 SNPs was selected in order to classify the 71 isolates used in the study into 25 SNP profiles. A minimum set of 5 SNPs was required to allocate the isolates into the six major clusters but the full set is required in the case of other *S.* Typhi populations (Octavia and Lan, 2008).

Pang et al. (2012) argued on the capacity of SNP typing to replace the traditional phage typing scheme for *S.* Typhimurium isolates due to the improved level of information that application of this technique offers. In that study, a total of 221 *S.* Typhimurium from 45 phage types were assayed using 44 SNPs and separated into 33 SNP profiles and four distinct phylogenetic groups. Moreover, SNP typing provided with phylogenetic information and thus the possible origin of 14 phage types could be assessed. Although it was not possible to differentiate between all phage types that were employed, the capacity of the approach has been exhibited.

Whole Genome Sequencing (WGS)

Whole Genome Sequencing (WGS) is a quite promising typing tool regarding epidemiological surveillance. However, current sequencing technology makes it at least unpractical.

WGS as a typing method for outbreak investigation of *S.* enterica strains was evaluated by Leekitcharoenphon et al. (2014). A collection of *S.* Typhimurium, *S.* Enteritidis and *S.* Derby isolates, collected from six different outbreaks, were sequenced and analyzed by a number of different bioinformatics approaches. The pan-genome approach, i.e. the approach based on the presence or absence of genes across the entire genomes was only successful in clustering strains according to their phage type and not according to the outbreak sources. The k-mer approach, i.e. the approach based on the frequency of defined k-mers across the genomes, although improved the concordance, i.e. the clustering of the strains according to the outbreak, when k was equal or bigger that 30, it failed to provide with complete outbreak typing. The latter was obtained for *S.* Typhimurium strains when they were clustered according to the nucleotide differences compared to a reference genome. This led to the conclusion that SNP distance within outbreak strains was less than between outbreaks. This conclusion was supported when strains of *S.* Montevideo from other outbreaks were analyzed.

However, this was not the case for *S.* Agona strains in which the number of SNPs within outbreak strains was higher than the respective between them. Moreover, the importance of the reference genome selection was also highlighted since it failed to classify the *S.* Enteriditis and *S.* Derby strains according to their outbreak source.

Clustered Regularly Interspaced Short Palindrome Repeats (CRISPRs)

The presence of short palindromic repeats of 25-50 nucleotides long that are interspaced by unique sequences of similar size has been recognized in both bacterial and archaeal genomes (van Belkum et al., 1998). These arrays of repeats are formed by embedding exogenously acquired nucleic acids into endogenous ones. CRISPR-associated proteins (Cas) identify these foreign molecules and incorporate them into the CRISPR locus. The biological significance of this mechanism has not yet elucidated, however the gain of some degree of immunity to the harmful foreign elements seems to be a reasonable explanation (Barrangou et al., 2007; Marraffini and Sontheimer, 2010; Bhaya et al., 2011; Horvath and Barrangou, 2010; van der Oost, 2009). The applicability of these arrays for microbial subtyping has been evaluated for several lactic acid bacteria (Horvath et al., 2009), *Yersinia pestis* (Pourcel et al., 2005), *Campylobacter jejuni* (Schouls et al., 2003), *Streptococcus pyogenes* (Hoe et al., 1999)

As far as foodbone *Salmonella* was concerned, a few studies have evaluated the utility of CRISPRs as a subtyping technique.

The genome analysis and CRISPR typing of *S.* Virchow isolates was reported by Bachmann et al. (2014). A total of 15 isolates from 7 different outbreaks had their CRISPR-1 region sequenced; they were separated into three allelic types failing thus to allocate the isolates according to the outbreaks.

The genetic diversity and evolutionary history of CRIPSR loci and Cas genes of a total of 427 *S. enterica* isolates belonging to 64 serovars was assessed by Pettengill et al. (2014). Moreover, the performance of CRISPR loci for *Salmonella* subtyping compared to whole genome and MLST typing schemes was evaluated. It was concluded that WGS provided with better discriminatory power than MLST and CRISPR assays. Regarding the latter, although allocated correctly a significant number of isolates within the respective serovar cluster, it failed to differentiate between outbreak related and unrelated isolates of *S.* Montevideo and *S.* Agona.

On the contrary, Fabre et al. (2012) studied the CRISPR polymorphisms in 783 reference strains and isolates from 130 different serotypes and concluded that the level of variation was adequate to differentiate between the most common serotypes involved in human infections and even between MLST groups within polyphyletic serotypes. These spacers variations increased the discrimination capacity to the level of PFGE.

ACKNOWLEDGMENTS

The authors gratefully acknowledge funding from the European Union Seventh Framework Programme (FP7/2007-2013) under grant agreement n. 289719 (Project QUAFETY).

REFERENCES

Achtman, M; Wain, J; Weill, FX; Nair, S; Zhou, Z; Sangal, V; Krauland, MG; Hale, JL; Harbottle, H; Uesbeck, A; Dougan, G; Harrison, LH; Brisse, S. Multilocus sequence typing as a replacement for serotyping in *Salmonella* enterica. *PLoS Pathog.,* 2012, 8, e1002776.

Allix, C; Supply, P; Fauville-Dufaux, M. Utility of fast mycobacterial interspersed repetitive unit-variable number tandem repeat genotyping in clinical mycobacteriological analysis. *Clin. Infect. Dis*., 2004, 39, 783-789.

Almeida, F; Pitondo-Silva, A; Oliveira, MA; Falcao, JP. Molecular epidemiology and virulence markers of *Salmonella* Infantis isolated over 25 years in Sao Paulo State, Brazil. *Infect. Gen. Evol*., 2013, 19, 145-151.

Bachmann, NL; Petty, NK; Ben Zakour, NL; Szubert, JM; Savill J; Beatson, SA. Genome analysis and CRISPR typing of *Salmonella* enterica serovar Virchow. *BMC Genomics*, 2014, 15, 389.

Barrangou, R; Fremaux, C; Deveau, H; Richards, M; Boyaval, P; Moineau, S; Romero, DA; Horvath, P. CRISPR provides acquired resistance against viruses in prokaryotes. *Science* 2007, 315, 1709-1712.

Barrett, TJ; Gerne-Smidt, P; Swaminathan, B. Interpretation of pulsed-field gel electrophoresis patterns in foodborne disease investigations and surveillance. *Foodborne Pathog. Dis*., 2006, 3, 20-31.

Beaubrun, JJG; Cheng, CM; Chen, KS; Ewing, L; Wang, H; Agpaoa, MC; Huang, MCJ; Dickey, E; Du, JM; Williams-Hill, DM; Hamilton, B; Micallef, SA; Rosenberg Goldstein, RE; George, A; Joseph, SW; Sapkota, AR; Jacobson, AP; Tall, BD; Kothary, MH; Dudley, K; Hanes, DE. The evaluation of a PCR-based method for identification of *Salmonella enterica* serotypes from environmental samples and various food matrices. *Food Microbiol*., 2012, 31, 199-209.

Bell, RL; Gonzalez-Escalona, N; Stones, R; Brown, EW. Phylogenetic evaluation of the 'Typhimurium' complex of *Salmonella* strains using a seven-gene multi-locus sequence analysis. *Infect. Gen. Evol*., 2011, 11, 83-91.

Beranek, A; Mikula, C; Rabold, P; Arnhold, D; Berghold, C; Lederer, I; Allerberger, F; Kornschober, C. Multiple-locus variable-number tandem repeat analysis for subtyping of *Salmonella enterica* subsp. *enterica* serovar Enteritidis. *Int. J. Med. Microbiol*., 2009, 299, 43-51.

Bergamini, F; Iori, A; Massi, P; Pongolini, S. Multilocus variable-number of tandem-repeats analysis of *Salmonella enterica* serotype Gallinarum and comparison with pulsed-field gel electrophoresis genotyping. *Vet. Microbiol*., 2011, 149, 430-436.

Bhaya, D; Davison, M; Barrangou, R. CRISPR-Cas systems in bacteria and archaea: versatile small RNAs for adaptive defense and regulation. *Ann. Rev. Genet.,* 2011, 45, 273-297.

Boxrud, D; Pederson-Gulrud, K; Wotton, J; Medus, C; Lyszkowicz, E; Besser, J; Bartkus, JM. Comparison of Multiple-Locus Variable-Number Tandem Repeat Analysis, Pulsed-Field Gel Electrophoresis, and Phage Typing for subtype analysis of *Salmonella enterica* serotype Enteritidis. *J. Clin. Microbiol*., 2007, 45, 536-543.

CDC. Pathogens causing US foodborne illnesses, hospitalizations, and deaths, 2000–2008. 2012. Available in http://www.cdc.gov/salmonella/ Accessed September 2014.

Chiou, CS; Hung, CS; Torpdahl, M; Watanabe, H; Tung, SK; Terajima, J; Liang, SY; Wang, YW. Development and evaluation of multilocus variable number tandem repeat analysis for fine typing and phylogenetic analysis of *Salmonella enterica* serovar Typhimurium. Int. *J. Food Microbiol.*, 2010, 142, 67-73.

Cho, S; Boxrud, DJ; Bartkus, JM; Whittam, TS; Saeed, M. Multiple locus variable-number tandem repeat analysis of *Salmonella* Enteritidis isolates from human and non-human sources using a single multiplex PCR. *FEMS Microbiol. Lett.*, 2007, 275, 16-23.

Cho, S; Whittam, TS; Boxrud, DJ; Bartkus JM; Saeed, AM. Allele distribution and genetic diversity of VNTR loci in *Salmonella enterica* serotype Enteritidis isolates from different sources. *BMC Microbiol.,* 2008, 8, 146.

Davis, MA; Baker, KNK; Call, DR; Warnick, LD; Soyer, Y; Wiedmann, M; Grohn, Y; McDonough, PL; Hancock, DD; Besser, TE. Multilocus Variable-Number Tandem-Repeat method for typing *Salmonella enterica* serovar Newport. *J. Clin. Microbiol.*, 2009, 47, 1934-1938.

Doulgeraki, AI; Paramithiotis, S; Kagkli, DM; Nychas, GJE. Lactic acid bacteria population dynamics during minced beef storage under aerobic or modified atmosphere packaging conditions. *Food Microbiol.,* 2010, 27, 1028-1034.

Doulgeraki, AI; Paramithiotis, S; Nychas, GJE. Development of *Enterobacteriaceae* community during storage of minced beef under aerobic or modified atmosphere packaging conditions. *Int. J. Food Microbiol.,* 2011, 145, 77-83.

Ducey, TF; Page, B; Usgaard, T; Borucki, MK; Pupedis, K; Ward, TJ. A single-nucleotide-polymorphism-based multilocus genotyping assay for subtyping lineage I isolates of *Listeria monocytogenes*. *Appl. Environ. Microbiol.*, 2007, 73, 133-147.

Fabre, L; Zhang, J; Guigon, G; Le Hello, S; Guibert, V; Accou-Demartin, M; de Romans, S; Lim, C; Roux, C; Passet, V; Diancourt, L; Guibourdenche, M; Issenhuth-Jeanjean, S; Achtman, M; Brisse, S; Sola, C; Weill, FX. CRISPR typing and subtyping for improved laboratory surveillance of *Salmonella* infections. *PLoS ONE*, 2012, 7, e36995.

Fakhr, MK; Nolan, LK; Logue, CM. Multilocus sequence typing lacks the discriminatory ability of Pulsed-Field Gel Electrophoresis for typing *Salmonella enterica* serovar Typhimurium. *J. Clin. Microbiol.*, 2005, 43, 2215-2219.

Farlow, J; Smith, KL; Wong, J; Abrams, M; Lytle, M; Keim, P. *Francisella tularensis* strain typing using multiple-locus, variable-number tandem repeat analysis. *J. Clin. Microbiol.*, 2001, 39, 3186-3192.

Feil, EJ; Cooper, JE; Grundmann, H; Robinson, DA; Enright, MC; Berendt, T; Peacock, SJ; Smith, JM; Murphy, M; Spratt, BG; Moore, CE; Day, NPJ. How clonal is *Staphylococcus aureus*? *J. Bacteriol.*, 2003, 185, 3307-3316.

Guedda, I; Taminiau, B; Ferjani, A; Boukadida, J; Bertrand, S; Daube, G. Antimicrobial and molecular analysis of *Salmonella* serovar Livingstone strains isolated from humans in Tunisia and Belgium. *J. Infect. Dev. Ctries.,* 2014, 8, 973-980.

Harbottle, H; White, DG; McDermott, PF; Walker, RD; Zhao S. Comparison of Multilocus Sequence Typing, Pulsed-Field Gel Electrophoresis, and Antimicrobial Susceptibility Typing for characterization of *Salmonella enterica* serotype Newport isolates. *J. Clin. Microbiol.*, 2006, 44, 2449-2457.

Hoe, N; Nakashima, K; Grigsby, D; Pan, X; Dou, SJ; Naidich, S; Garcia, M; Kahn, E; Bergmire-Sweat, D; Musser, JM. Rapid molecular genetic subtyping of serotype M1 group A *Streptococcus* strains. *Emerg. Infect. Dis.*, 1999, 43, 254-263.

Hopkins, KL; Maguire, C; Best, E; Liebana, E; Threlfall EJ., Stability of Multiple-Locus Variable-Number Tandem Repeats in *Salmonella enterica* serovar Typhimurium. *J. Clin. Microbiol.*, 2007, 45, 3058-3061.

Horvath, P; Barrangou, R. CRISPR/Cas, the immune system of bacteria and archaea. *Science* 2010, 327, 167-170.

Horvath, P; Coute-Monvoisin, AC; Romero, DA; Boyaval, P; Fremaux, C; Barrangou, R. Comparative analysis of CRISPR loci in lactic acid bacteria genomes. *Int. J. Food Microbiol.,* 2009, 131, 62-70.

Iredell, J; Blanckenberg, D; Arvand, M; Grauling, S; Feil, EJ; Birtles, RJ. Characterization of the natural population of *Bartonella henselae* by multilocus sequence typing. *J. Clin. Microbiol.*, 2003, 41, 5071-5079.

Jeon, SM; Lim, NR; Kwon, SJ; Shim, TS; Park, MS; Kim, BJ; Kim, SH. Analysis of species and intra-species associations between the *Mycobacterium abscessus* complex strains using pulsed-field gel electrophoresis (PFGE) and multi-locus sequence typing (MLST). *J. Microbiol. Meth.*, 2014, 104, 19-25.

Johansson, A; Farlow, J; Larsson, P; Dukerich, M; Chambers, E; Bystrom, M; Fox, J; Chu, M; Forsman, M; Sjostedt, A; Keim, P. Worldwide genetic relationships among *Francisella tularensis* isolates determined by multiple-locus variable-number tandem repeat analysis. *J. Bacteriol.*, 2004, 186, 5808-5818.

Keim, P; Price, LB; Klevytska, AM; Smith, KL; Schupp, JM; Okinaka, R; Jackson, PJ; Hugh-Jones, ME. Multiple-locus variable-number tandem repeat analysis reveals genetic relationships within *Bacillus anthracis*. *J. Bacteriol.*, 2000, 182, 2928-2936.

Kim, S; Frye, JG; Hu, J; Fedorka-Cray, PJ; Gautom, R; Boyle, DS. Multiplex PCR-Based Method for identification of common clinical serotypes of *Salmonella enterica* subsp. *enterica*. *J. Clin. Microbiol.*, 2006, 44, 3608-3615.

King, SJ; Leigh, JA; Heath, PJ; Luque, I; Tarradas, C; Dowson, CG; Whatmore, AM. Development of a multilocus sequence typing scheme for the pig pathogen *Streptococcus suis*: identification of virulent clones and potential capsular serotype exchange. *J. Clin. Microbiol.*, 2002, 40, 3671-3680.

Klevytska, AM; Price, LB; Schupp, JM; Worsham, PL; Wong, J; Keim, P. Identification and characterization of variable-number tandem repeats in the *Yersinia pestis* genome. *J. Clin. Microbiol.*, 2001, 39, 3179-3185.

Kotetishvili, M; Stine, OC; Chen, Y; Kreger, A; Sulakvelidze, A; Sozhamannan, S; Morris Jr, JG. Multilocus sequence typing has better discriminatory ability for typing Vibrio cholerae than does pulsed-field gel electrophoresis and provides a measure of phylogenetic relatedness. *J. Clin. Microbiol.*, 2003, 41, 2191-2196.

Kotetishvili, M; Stine, OC; Kreger, A; Morris Jr, JG; Sulakvelidze, A. Multilocus Sequence typing for characterization of clinical and environmental *Salmonella* strains. *J. Clin. Microbiol.*, 2002, 40, 1626-1635.

Ktari, S; Arlet, G; Mnif, B; Gautier, V; Mahjoubi, F; Ben Jmeaa, M; Bouaziz, M; Hammami, A. Emergence of multidrug-resistant Klebsiella pneumoniae isolates producing *VIM*-4 metallo-beta-lactamase, *CTX-M*-15 extended-spectrum beta-lactamase, and *CMY*-4 AmpC beta-lactamase in a Tunisian university hospital. *Antimicrob. Agents Chemother.*, 2006, 50, 4198-4201.

Ktari, S; Arlet, G; Verdet, C; Jaoua, S; Kachrid, A; Ben Redjeb, S; Mahjoubi-Rhimi, F; Hammami, A. Molecular epidemiology and genetic environment of acquired *bla ACC*-1 in *Salmonella enterica* serotype Livingstone causing a large nosocomial outbreak in Tunisia. *Microb. Drug Resist.*, 2009, 15, 279-286.

Leader, BT; Frye, JG; Hu, J; Fedorka-Cray, PJ; Boyle, DS. High-Throughput molecular determination of Salmonella enterica serovars by use of multiplex PCR and capillary electrophoresis analysis. *J. Clin. Microbiol.*, 2009. 47, 1290-1299.

Leekitcharoenphon, P; Nielsen, EM; Kaas, RS; Lund, O; Aarestrup, FM. Evaluation of whole genome sequencing for outbreak detection of *Salmonella enterica*. *PLoS ONE*, 2014, 9, e87991.

Lettini, AA; Saccardin, C; Ramon, E; Longo, A; Cortini, E; Dalla Pozza, MC; Barco, L; Guerra, B; Luzzi, I; Ricci, A. Characterization of an unusual *Salmonella* phage type DT7a and report of a foodborne outbreak of salmonellosis *Int. J. Food Microbiol.*, 2014, 189, 11-17.

Liang, SY; Watanabe, H; Terajima, J; Li, CC; Liao, JC; Tung, SK; Chiou, CS. Multilocus variable-number tandem repeat analysis for molecular typing of *Shigella sonnei*. *J. Clin. Microbiol.*, 2007, 45, 3574-3580.

Lindstedt, BA; Heir, E; Gjernes, E; Kapperud G. DNA Fingerprinting of *Salmonella enterica* subsp. *enterica* serovar Typhimurium with emphasis on phage type DT104 based on variable number of tandem repeat loci. *J. Clin. Microbiol.*, 2003a, 41, 1469-1479.

Lindstedt, BA; Heir, E; Gjernes, E; Vardund, T; Kapperud, G. DNA fingerprinting of Shiga-toxin producing *Escherichia coli* O157 based on multiple-locus variable-number tandem-repeats analysis (MLVA). *Ann. Clin. Microbiol. Antimicrob,*. 2003b, 2, 12.

Lindstedt, BA; Vardund, T; Aas, L; Kapperud, G. Multiple-locus variable-number tandem-repeats analysis of *Salmonella enterica* subsp. *enterica* serovar Typhimurium using PCR multiplexing and multicolor capillary electrophoresis. *J. Microbiol. Meth.*, 2004, 59, 163-172.

Liu, B; Zhang, L; Zhu, X; Shi, C; Chen, J; Liu, W; He, X; Shi, X. PCR identification of *Salmonella* serogroups based on specific targets obtained by comparative genomics. *Int. J. Food Microbiol.*, 2011a, 144, 511-518.

Liu, F; Barrangou, R; Gerner-Smidt, P; Ribot, EM; Knabel, SJ; Dudley, EG. Novel virulence gene and Clustered Regularly Interspaced Short Palindromic Repeat (CRISPR) multilocus sequence Typing scheme for subtyping of the major serovars of *Salmonella enterica* subsp. *enterica*. *Appl. Environ. Microbiol.*, 2011b, 77, 1946-1956.

Liu, F; Kariyawasam, S; Jayarao, BM; Barrangou, R; Gerner-Smidt, P; Ribot, EM; Knabel, SJ; Dudley, EG. Subtyping *Salmonella enterica* Serovar Enteritidis isolates from different sources by using sequence typing based on virulence genes and Clustered Regularly Interspaced Short Palindromic Repeats (CRISPRs). *Appl. Environ. Microbiol.*, 2011c, 77, 4520-4526.

Liu, W; Liu, B; Zhu, X; Yu, S; Shi, X. Diversity of Salmonella isolates using serotyping and multilocus sequence typing. *Food Microbiol.*, 2011d, 28, 1182-1189.

Liu, Y; Lee, MA; Ooi, EE; Mavis, Y; Tan, AL; Quek, HH. Molecular typing of *Salmonella enterica* Serovar Typhi isolates from various countries in Asia by a multiplex PCR assay on Variable-Number Tandem Repeats. *J. Clin. Microbiol.*, 2003, 41, 4388-4394.

Maiden, MCJ; Bygraves, JA; Feil, E; Morelli, G; Russell, JE; Urwin, R; Zhang, Q; Zhou, J; Zurth, K; Caugant, DA; Feavers, IM; Achtman, M; Spratt, BG. Multilocus sequence typing: A portable approach to the identification of clones within populations of pathogenic microorganisms *Proc. Natl. Acad. Sci. USA*, 1988, 95, 3140-3145.

Malorny, B; Huehn, S; Dieckmann, R; Krämer, N; Helmuth, R. Polymerase chain reaction for the rapid detection and serovar identification of *Salmonella* in food and feeding stuff. *Food Analyt. Meth.*, 2009, 2, 81-95.

Malorny, B; Junker, E; Helmuth, R. Multi-locus variable-number tandem repeat analysis for outbreak studies of *Salmonella enterica* serotype Enteritidis. *BMC Microbiol.,* 2008, 8, 84.

Manning, SD; Motiwala, AS; Springman, AC; Qi, W; Lacher, DW; Ouellette, LM; Mladonicky, JM; Somsel, P; Rudrik, JT; Dietrich, SE; Zhang, W; Swaminathan, B; Alland, D; Whittam, TS. Variation in virulence among clades of *Escherichia coli* O157:H7 associated with disease outbreaks. *Proc. Natl. Acad. Sci. USA,* 2008, 105, 4868-4873.

Marraffini, LA; Sontheimer, EJ. CRISPR interference: RNA-directed adaptive immunity in bacteria and archaea. *Nat. Rev. Genet.,* 2010, 11, 181-190.

McClelland, M; Sanderson, KE; Clifton, SW; Latreille, P; Porwollik, S; Sabo, A; Meyer, R; Bieri, T; Ozersky, P; McLellan, M; Harkins, CR; Wang, C; Nguyen, C; Berghoff, A; Elliott, G; Kohlberg, S; Strong, C; Du, F; Carter, J; Kremizki, C; Layman, D; Leonard, S; Sun, H; Fulton, L; Nash, W; Miner, T; Minx, P; Delehaunty, K; Fronick, C; Magrini, V; Nhan, M; Warren, W; Florea, L; Spieth, J; Wilson, RK. Comparison of genome degradation in Paratyphi A and Typhi, human-restricted serovars of *Salmonella enterica* that cause typhoid. *Nat. Genet.*, 2004, 36, 1268-1274.

McTavish, SM; Pope, CE; Nicol, C; Campbell, D; French, N; Carter, PE. Multilocus sequence typing of *Campylobacter jejuni*, and the correlation between clonal complex and pulsed-field gel electrophoresis macrorestriction profile. *FEMS Microbiol. Lett.*, 2009, 298, 149-156.

Noller, AC; McEllistrem, MC; Pacheco, AG; Boxrud, DJ; Harrison, LH. Multilocus variable-number tandem repeat analysis distinguishes outbreak and sporadic *Escherichia coli* O157:H7 isolates. *J. Clin. Microbiol.*, 2003, 41, 5389-5397.

Octavia S; Lan, R. Single nucleotide polymorphism typing and genetic relationships of *Salmonella enterica* serovar Typhi isolates. *J. Clin. Microbiol.*, 2007, 45, 3795-3801.

Octavia, S; Lan, R. Multiple-Locus Variable-Number Tandem-Repeat Analysis of *Salmonella enterica* serovar Typhi. *J. Clin. Microbiol.*, 2009, 47, 2369-2376.

Octavia, S; Lan, R. Single Nucleotide Polymorphism typing of global *Salmonella enterica* Serovar Typhi isolates by use of a hairpin primer Real-Time PCR assay. *J. Clin. Microbiol.*, 2010, 48, 3504-3509.

Octavia, S; Lan, R. Single-Nucleotide-Polymorphism typing and genetic relationships of *Salmonella enterica* serovar Typhi isolates. *J. Clin. Microbiol.*, 2007, 45, 3795-3801.

Onteniente, L; Brisse, S; Tassios, PT; Vergnaud, G. Evaluation of the polymorphisms associated with tandem repeats for *Pseudomonas aeruginosa* strain typing. *J. Clin. Microbiol.*, 2003, 41, 4991-4997.

Overduin, P; Schouls, L; Roholl, P; van der Zanden, A; Mahmmod, N; Herrewegh, A; van Soolingen, D. Use of multilocus variable-number tandem-repeat analysis for typing *Mycobacterium avium* subsp. *paratuberculosis*. *J. Clin. Microbiol.*, 2004, 42, 5022-5028.

Pang, S; Octavia, S; Reeves, PR; Wang, Q; Gilbert, GL; Sintchenko, V; Lan, R. Genetic relationships of phage types and Single Nucleotide Polymorphism typing of *Salmonella enterica* serovar Typhimurium. *J. Clin. Microbiol.,* 2012, 50, 727-734.

Paramithiotis, S; Kagkli, DM; Blana, VA; Nychas, GJE; Drosinos, EH. Identification and characterization of *Enterococcus* spp. in Greek spontaneous sausage fermentation. *J. Food Prot.*, 2008, 71, 1244-1247.

Parisi, A; Latorre, L; Normanno, G; Miccolupo, A; Fraccalvieri, R; Lorusso, V; Santagada, G. Amplified fragment length polymorphism and multi-locus sequence typing for high resolution genotyping of *Listeria monocytogenes* from foods and the environment. *Food Microbiol.*, 2010, 27, 101-108.

Pettengill, JB; Timme, RE; Barrangou, R; Toro, M; Allard, MW; Strain, E; Musser SM; Brown, EW. The evolutionary history and diagnostic utility of the CRISPR-Cas system within *Salmonella enterica* ssp. *enterica*. *PeerJ*, 2014, 2, e340

Pourcel, C; Salvignol, G; Vergnaud, G. CRISPR elements in *Yersinia pestis* acquire new repeats by preferential uptake of bacteriophage DNA, and provide additional tools for evolutionary studies. *Microbiol.*, 2005, 151, 653-663.

Ramisse, V; Houssu, P; Hernandez, E; Denoeud, F; Hilaire, V; Lisanti, O; Ramisse, F; Cavallo, JD; Vergnaud, G. Variable number of tandem repeats in *Salmonella enterica* subsp. *enterica* for typing purposes. *J. Clin. Microbiol.*, 2004, 42, 5722-5730.

Revazishvili, T; Kotetishvili, M; Stine, OC; Kreger, AS; Morris Jr, JG; Sulakvelidze, A. Comparative analysis of multilocus sequence typing and pulsed-field gel electrophoresis for characterizing *Listeria monocytogenes* strains isolated from environmental and clinical sources. *J. Clin. Microbiol.*, 2004, 42, 276-285.

Ross, IL; Parkinson, IH; Heuzenroeder, MW. The use of MAPLT and MLVA analyses of phenotypically closely related isolates of *Salmonella enterica* serovar Typhimurium. *Int. J. Med. Microbiol.*, 2009, 299, 37-41.

Schouls, LM; Reulen, S; Duim, B; Wagenaar, JA; Willems, RJ; Dingle, KE; Colles, FM; Van Embden, JD. Comparative genotyping of *Campylobacter jejuni* by amplified fragment length polymorphism, multilocus sequence typing, and short repeat sequencing: strain diversity, host range, and recombination. *J. Clin. Microbiol.*, 2003, 41, 15–26.

Schouls, LM; van der Ende, A; van de Pol, I; Schot, C; Spanjaard, L; Vauterin, P; Wilderbeek, D; Witteveen, S. Increase in genetic diversity of *Haemophilus influenzae* serotype b (Hib) strains after introduction of Hib vaccination in The Netherlands. *J. Clin. Microbiol.* 2005, 43, 2741-2749.

Schouls, LM; van der Heide, HG; Vauterin, L; Vauterin, P; Mooi, FR. Multiple-locus variable-number tandem repeat analysis of Dutch *Bordetella pertussis* strains reveals rapid genetic changes with clonal expansion during the late 1990s. *J. Bacteriol.*, 2004, 186, 5496-5505.

Shariat, N; DiMarzio, MJ; Yin, S; Dettinger, L; Sandt, CH; Lute, JR; Barrangou, R; Dudley, EG. The combination of CRISPR-MVLST and PFGE provides increased discriminatory power for differentiating human clinical isolates of *Salmonella enterica* subsp. *enterica* serovar Enteritidis. *Food Microbiol.*, 2013c, 34, 164-173.

Shariat, N; Kirchner, MK; Sandt, CH; Trees, E; Barrangou, R; Dudley, EG. Subtyping of *Salmonella enterica* serovar Newport outbreak isolates by CRISPR-MVLST and determination of the relationship between CRISPR-MVLST and PFGE results. *J. Clin. Microbiol.* 2013b, 51, 2328-2336.

Shariat, N; Sandt, CH; DiMarzio, MJ; Barrangou, R; Dudley, EG. CRISPR-MVLST subtyping of *Salmonella enterica* subsp. *enterica* serovars Typhimurium and Heidelberg and application in identifying outbreak isolates. *BMC Microbiol.*, 2013a, 13, 254.

Singh, P; Foley, SL; Nayak, R; Kwon, YM. Multilocus sequence typing of *Salmonella* strains by high-throughput sequencing of selectively amplified target genes. *J. Microbiol. Meth.*, 2012, 88, 127-133.

Sukhnanand, S; Alcaine, S; Warnick, LD; Su, WL; Hof, J; Pat, M; Craver, J; McDonough, P; Boor, KJ; Wiedmann M. DNA sequence-based subtyping and evolutionary analysis of selected *Salmonella enterica* serotypes. *J. Clin. Microbiol.*, 2005, 43, 3688-3698.

Svraka, S; Toman, R; Skultety, L; Slaba, K; Homan, WL. Establishment of a genotyping scheme for *Coxiella burnetii*. *FEMS Microbiol. Lett.*, 2006, 254, 268-274.

Swaminathan, B; Barrett, TJ; Hunter, SB; Tauxe, RV. PulseNet: the molecular subtyping network for foodborne bacterial disease surveillance, United States. *Emerg. Infect. Dis.*, 2001, 7, 382-389.

Tankouo-Sandjong, B; Sessitsch, A; Liebana, E; Kornschober, C; Allerberger, F; Hachler, H; Bodrossy, L. MLST-v, multilocus sequence typing based on virulence genes, for molecular typing of *Salmonella enterica* subsp. *enterica* serovars. *J. Microbiol. Meth.*, 2007, 69, 23-36.

Tenover, FC; Arbeit, RD; Goering, RV; Mickelsen, PA; Murray, BE; Persing, DH; Swaminathan, B. Interpreting chromosomal DNA restriction patterns produced by Pulsed-Field Gel Electrophoresis: criteria for bacterial strain typing. *J Clin. Microbiol.*, 1995, 33, 2233-2239.

Tien, YY; Ushijima, H; Mizuguchi, M; Liang, SY; Chiou, CS. Use of multilocus variable-number tandem repeat analysis in molecular subtyping of *Salmonella enterica* serovar Typhi isolates. *J. Med. Microbiol.*, 2012, 61, 223-232.

Tien, YY; Wang, YW; Tung, SK; Liang, SY; Chiou, CS. Comparison of multilocus variable-number tandem repeat analysis and pulsed-field gel electrophoresis in molecular subtyping of *Salmonella enterica* serovars Paratyphi A. *Diagn. Micr. Infect. Dis.*, 2011, 69, 1-6.

Titze-de-Almeida, R; Willems, RJ; Top, J; Rodrigues, IP; Ferreira, RF; Boelens, H; Brandileone, MC; Zanella, RC; Felipe, MS; van Belkum, A. Multilocus variable-number tandem-repeat polymorphism among Brazilian *Enterococcus faecalis* strains. *J. Clin. Microbiol.*, 2004, 42, 4879-4881.

Toboldt, A; Tietze, E; Helmuth, R; Junker, E; Fruth, A; Malorny, B. Molecular epidemiology of *Salmonella enterica* serovar Kottbus isolated in Germany from humans, food and animals. *Vet. Microbiol.*, 2014, 170, 97-108.

Torpdahl, M; Skov, MN; Sandvang, D; Baggesen, DL. Genotypic characterization of *Salmonella* by multilocus sequence typing, pulsed-field gel electrophoresis and amplified fragment length polymorphism. *J. Microbiol. Meth.*, 2005, 63, 173-184.

van Belkum, A; Scherer, S; van Alphen, L; Verbrugh, H. Short-sequence DNA repeats in prokaryotic genomes. *Microbiol. Mol. Biol. Rev.*, 1998, 43, 275-293.

van Belkum, A; Tassios, PT; Dijkshoorn, L; Haeggman, S; Cookson, B; Fry, NK; Fussing, V; Green, J; Feil, E; Gerner-Smidt, P; Brisse, S; Struelens, M. Guidelines for the validation and application of typing methods for use in bacterial epidemiology. *Clin. Microbiol. Infect. Dis.*, 2007, 13, 1-46.

van der Oost, J; Jore, MM; Westra, ER; Lundgren, M; Brouns, SJ. CRISPR-based adaptive and heritable immunity in prokaryotes. *Trends Biochem. Sci.*, 2009, 34, 401-407.

van Loo, IHM; Heuvelman, KJ; King, AJ; Mooi, FR. Multilocus sequence typing of *Bordetella pertussis* based on surface protein genes. *J. Clin. Microbiol.*, 2002, 40, 1994-2001.

Ward, TJ; Ducey ,TF; Usgaard, T; Dunn, KA; Bielawski, JP. Multilocus genotyping assays for single nucleotide polymorphism-based subtyping of *Listeria monocytogenes* isolates. *Appl. Environ. Microbiol.*, 2008, 74, 7629-7642.

Yang, M; Shao, S; Xiao, J; Wang, Q; Zhang, Y. Phylogenetic investigation of *Edwardsiella tarda* with multilocus sequence typing (MLST) and pulsed field gel electrophoresis (PFGE) typing methods. *Aquaculture*, 2013, 410-411, 79-85.

Yazdankhah, SP; Lindstedt, BA; Caugant, DA. Use of variable-umber tandem repeats to examine genetic diversity of *Neisseria meningitidis*. *J. Clin. Microbiol.* 2005, 43, 1699-1705.

Young, CC; Ross, IL; Heuzenroeder, MW. New methodology for differentiation and typing of closely related *Salmonella enterica* serovar Heidelberg isolates. *Curr. Microbiol.*, 2012, 65, 481-487.

Zhang, W; Qi, W; Albert, TJ; Motiwala, AS; Alland, D; Hyytia-Trees, EK; Ribot, EM; Fields, PI; Whittam, TS; Swaminathan, B. Probing genomic diversity and evolution of *Escherichia coli* O157 by single nucleotide polymorphisms. *Genome Res.*, 2006, 16, 757-767.

In: Salmonella
Editor: Christopher B. Hackett
ISBN: 978-1-63463-651-3

Chapter 9

SALMONELLA: RESISTANCE FOR MANY YEARS UNDER STRESSFUL CONDITIONS AND POSSIBILITY OF RESUSCITATION

Dhiaf Amel* and Bakhrouf Amina
Laboratoire d'Analyse et de Contrôle des Polluants Chimiques et Microbiologiques de l'Environnement, Tunisia

ABSTRACT

Salmonella is a gram-negative microorganism, generally dispersed in nature and habitually found in the intestinal tract of animals and humans. *Salmonella* reveals a considerable aptitude to adapt to physical and chemical stresses in the environment. Survival mechanisms are stimulated following the detection of environmental signals and provoke a complicated adaptive response that guides to a state of tolerance and thus survival below sub-optimal or even sub-lethal conditions. This microorganism can persist for long periods in the environment in a heavily stressed state known variously, and often contentiously, as viable but nonculturable.

Salmonella as many other bacteria resist to environmental stress by adopting the viable but nonculturable phenotype (VBNC). VBNC bacteria have been exemplified as organisms that fail to grow and develop colonies on routine bacteriological media, but which accrue to be considered alive since of their aptitude of metabolic activity and preserve virulence.

Is it convincing that the VBNC phenotype can revert to a culturable state, and vice versa.

This chapter presents and evaluates the hypotheses concerning the resistance of VBNC forms of *Salmonella* under extended period of time of stressful conditions.

Keywords: *Salmonella*, extremely stressing conditions, viable but non culturable state, resuscitation, many years

---

* Email: ameldhiaf@yahoo.fr.

Introduction

Salmonella is a gram-negative microorganism, considered the second mostly major bacterial cause of foodborne gastroenteritis worldwide. *Salmonella* pathogens may frequently spread through the feces of wildlife and domestic animals, infected water, poor fertilization methods, and additional agricultural practices. *Salmonella* can develop and endure in many different foods. (Andrews and Baumler 2005). Once the environmental conditions menace their survival or stop them from living in finest conditions, the cells are expressed as stressed (Dhiaf and Bakhrouf 2007). This concept of stress plays an essential role in the survival of microorganisms in the environment. Nevertheless, between the unstressed state and death, diverse physiological states have been described (Steinert et al., 1997, Votyakova et al., 1995). Bacteria in the VBNC state fail to grow on the routine bacteriological media on which they would normally grow and develop into colonies, but are alive and capable of renewed metabolic activity (Oliver 1995).

Salmonella enter the viable but nonculturable state (VBNC) when they encounter environmental stresses, such as low temperatures, oligotrophic conditions, and biocides including heavy metals and ultraviolet radiation (Oliver 1995, Lindbäck et al., 2010). VBNC bacteria are organisms that fail to grow and develop colonies on media, but their metabolic activity capabilities indicate that they are still alive (Roszak, and Colwell 1987). Bacteria in the VBNC state show changes in cell physiology and morphology (Hammouda 2012).

Numerous bacteria persist for long period under environmental conditions appreciation to chronological modification in cellular physiology and morphology changes (Morita, 1993). Several studies proved that viable but not culturable VBNC state is an adaptive advance developed by microorganisms in response to stressful conditions. The VBNC state is definite as a form of dormancy of bacteria generated by hard environmental conditions (Olivier 2010). Bacteria that do not cultivated in solid culture media, but are still metabolically vigorous and able of causing infections, are assumed to be in a VBNC state. Habitual laboratory culture conditions and methods cannot assemble the requirements of VBNC organisms to recommence growth (Mizunoe et al., 2000). The VBNC state is defined as a state of dormancy triggered by harsh environmental conditions (Oliver 1993). This state is a extensive way of adaptation survival of stressed bacteria. VBNC bacteria show metabolic activity and respiration, but cannot be revealed as colony forming units by the usual plate counts and therefore continue concealed (Colwell et al., 1996). In fact, the cells can get their initial shape when the good conditions are returned (Dhiaf et al., 2004, 2008, 2010). While numerous investigations have reported on this aspect of the VBNC state, resuscitation has been most extensively studied (Oliver 1995).

Resistance of VBNC Salmonella for Many Years in Environmental Conditions

It had long been understood that a bacterial cell was dead when it was no longer able to grow on routine culture media. We currently discern that this supposition is crude, and that there are many circumstances where a cell loses culturability but remains viable and potentially able to resuscitate. Great amount of research has investigated the ability of VBNC forms to resist for many years in very hard environmental conditions and the ability of resuscitation, many reports investigated the fitness of VBNC resuscitation after many years of starvation in environmental microcosms. (Dhiaf et al., 2004, 2008, 2010, Ben Kahla-Nakbi et al., 2007, Elabid et al., 2012). Indeed, Dhiaf et al., (2007) observed that *Salmonella* Typhimurium persisted for an extended period of time (203 to 231 days), and could be detected on infected vegetative parts of the rosemary and barley plants even after desiccation. Colony morphology displayed a mucoid and rugose phenotype. Besides, Dhiaf et al., confirmed that VBNC *Salmonella* Typhimurium resisted in soil microcosms for 20 years, recovery was obtained after two months of incubation in nutrient broth and buffered peptone water. Dhiaf et al., (2001) also reported that 5 years seawater stressed *Salmonella* Typhimurium were recovered in nutrient broth and recuperate most of their biochemical characters after 5 months of incubation. In addition, following 13 years of starvation in natural soil microcosm, cells of *Salmonella* returned to cultivable state on TSA after 24 h of incubation in TSB. These cells showed a biochemical profile completely inactive on API 20E system. After 6 months of reviviscence in TSB, cells recuperated their initial characters (Bakhrouf et al., 2008).

Moreover, Dhiaf et al., intended the providence of *Salmonella* Typhimurium starved in seawater microcosms for seventeen years. The microcosms were prepared in such a way as to allow progressive evaporation of the water. Despite being introduced into the sterile seawater at very high concentrations, the *Salmonella* rapidly declined to levels undetectable by plate counts on nutrient agar.

After two years of starvation, about half of the seawater volume had evaporated from each microcosm, and salt crystals appeared.

Inoculation of the salty suspension did not result in any culturable strains in selective and non-selective media. Nevertheless, incubation of samples in nutrient-rich broth, without supplemental growth factors, allowed resuscitation of stressed cells, yielding colonies that remained viable for extended periods of time. After seventeen years no water remained in the microcosms, and only salt crystals were visible. Inoculation of salt in nutrient broth led to the formation of biofilms on the upper surfaces of culture tube walls after 48 hours. Colony morphology and biochemical profile of Api 20 E were acquired after three months of resuscitation in nutrient broth. Recovery of seventeen year stressed *Salmonella* was observed in mice model orally administrated by VBNC *Salmonella*. Resuscitated germ appeared in mice feces and organs (stomach, large and small intestines, liver and lungs) after only 24 hours of administration. However recovery was not observed following intraperitoneal injection into mice.

As well, Dhiaf et al., (2004) studied the fate of *Salmonella* Typhimurium and *Salmonella paratyphi* B strains exposed to sterile seawater respectively for 12 and 6 years. After starvation periods, no colonies were observed in selective and non-selective media. Following resuscitation in nutrient broth, starved cells developed rough colonies on nutrient agar medium.

Recovery was not obtained in non-embryonated eggs; however it was successfully acquired in yolk sac of embryonated chicken eggs.

Looking upon the literature cited above, we could consider that VBNC state of *Salmonella* could persist in environmental conditions for many years; resuscitation is obtained when the good conditions return.

Pathogenicity of Long Term Stressed Salmonella

Dhiaf et al., (2004a) revealed gastroenteritis infections due to *Salmonella* species origin of significant morbidity and mortality at three Tunisian hospitals from February through July 2000. VBNC forms of *Salmonella* existing in pasteurized powder milk were the main cause of this epidemiology. Dhiaf et al., (2004a) analyzed baby powder milk given for hospitalized baby; they confirmed the recovery of VBNC *Salmonella* in gastrointestinal tracts. Within one day, *Salmonella* infection cause septicemia and the germ appeared in several organs (stomach, intestines, liver and lungs). Dhiaf et al., (2004a) confirmed that babies' infections appeared from the first days of their feed by powder milk which confirm the rapid resuscitation and infective effect of VBNC forms in animal models. Two serotypes of *Salmonella* were isolated *Salmonella* spp. and *Salmonella* arizona. In addition, the study of Oliver and Bockian (1995) showed that injections of VBNC *Vibrio vulnificus* cells into mice killed the animals and concluded that VBNC V.vulnificus cells remain virulent, at least for some time after entry into the VBNC state, and are capable of causing fatal infection after recovery in vivo. Other authors have shown that VBNC *E. coli* cells retain pathogenicity, with cells being able to produce enterotoxin and maintain virulence plasmids (Pommepuy et al., 1996).

Dhiaf et al., (2004b) furthermore in other study proved the recovery of 12 years seawater starved *Salmonella* VBNC cells in embryonated eggs. The embryonated-egg model has been successfully used to recover VBNC *Legionella pneumophilacells* Hussong et al., 1987. This agree with the experimental results of Cappelier et al., (1999), in which passage in embryonated eggs is the preferable model for recovery of the VBNC stage of Campylobacter. The embryonated- egg model can be consideredan animal model in which the animals have reduced defences. Dhiaf et al., exhibited the resuscitation of *Salmonella* Typhimurium in mice model after seventeen years of starvation in seawater microcosms. Recovery of VBNC cells was not observed by intra-peritoneal injection into mice however it was successfully obtained when mice were orally ingested by resistant cells. Recovered *Salmonella* cells were clearly observed in different organs.

Morphological Alterations of Long Term Stressed Salmonella

The long-term survival of *Salmonella* in stressful conditions is accompanied by many morphological and metabolic modifications, among which the reduction of the cells size and the evolution to coccoid-shapes (Bakhrouf et al., Dhiaf et al.). According to Bakhrouf et al., (1989), *P. aeruginosa*, when incubated in sterile nutrient-free seawater, is able to modify some of its biochemical characters, such as gelatinase or urease activity. Numerous starved Gram negative bacteria generally recognized to have coccoid-shaped cells (Van Overbeek et al., 1995). The decrease of bacterial size through is a strategy of adaptation to minimize the requisition for the cells (Jiang and Chai, 1996). The reduction of cells size is the result of cytoplasmic retrenchment and volume (Huisman et al., 1996). Morphological changes of *E. coli* O157:H7 and S. typhimurium, when exposed to hydrostatic pressure and bacteriocin mixture, were observed by Kalchayanand et al. (2004) observed. Starved Vibrionaceae and Pseudomonadaceae represented many morphological and physiological changes (Roszak and Colwell, 1987). During stressful conditions, large alteration in membrane structure, protein composition, ribosomal content and possibly even DNA arrangement arise (Oliver 1995). In fact, decrease of cell size is a strategy of adaptation against hard conditions (Jiang and Chai 1996). The evolution towards this state make the bacterium resist for long period under starvation conditions.

Dhiaf et al., (2001) reported that strains of *Salmonella typhimurium* incubated during 5 years in seawater microcosms were cultivable after 24 h of incubation in nutrient broth and recuperated most of their biochemical characters after 5 months of resuscitation.

Villarino et al., (2000) showed that VBNC state could be metabolically reactivated. Ben Abdallah et al., (2007) reported that *Salmonella Typhimurium* losed some morphological characters after its incubation in seawater, indeed it can besides recover them when the favorable conditions are restored (Dhief et al., 2001).

Dhiaf et al., showed that long-term stressed *Salmonella* was revived after 24 hours of in nutrient broth, biofilm was clearly observed in the upper surface of tube walls. In fact bacterial biofilms are complex communities of microorganisms implanted in a self-produced matrix and sticked to inert or living surfaces (Costerton et al., 1999). Biofilms have been observed on a variety of surfaces and were considered to be the prevailing microbial lifestyle in most environments (Van Houdt et al., 2004). Biofilm is formed by polysaccharidic materials, while other polymers may be present. They are concerned in the defense of microbial cells. As well microorganisms which produce these exopolymers are more resistant to desiccation, predation and toxic chemicals (Ophir and Gutnick 1994). The attachment of bacteria, to glace, polystyrene or other surfaces is affected by various physicochemical and biological factors including bacterial surface hydrophobicity (Marshall 1971), surface appendages, extracellular polymeric substances (Characklis 1990), bacterial physiological state, electrolyte concentration in the medium (Marshall 1971) surface charge and swimming speed.

CONCLUSION

The present work showed that VBNC forms of *Salmonella* germ persists in environmental conditions for many years; resuscitation could be observed. The identification of stressed *Salmonella* by habitual laboratory technique is hard when the large morphological and/or biochemical characters have been lost. In fact, molecular biology methods are required for the identification *Salmonella* stressed for long-term in hostile environments.

REFERENCES

Amel Dhiaf, Fethi Ben Abd Allah and Amina Bakhrouf. Resuscitation of 20 year starved *Salmonella* in sea water and soil. 2010. Annals of Microbiology. vol. 60 page 157-160.

Amina Bakhrouf, Amel Dhiaf. *Salmonella* identification after incubation in soil 2002. Archive de l institut Pasteur Tunis. vol 1 page 35-41.

Amina Bakhrouf, Fethi Ben Abdallah, Rihab Lagha. Morphological changes of starved *Salmonella enterica* serovar *Agona* cells in soil after resuscitation. *Annals of Microbiology,* 58 (3) 521-525 (2008).

Amel BK-N, Amine B, Amina B. Survival of *Vibrio fluvialis* in seawater under starvation conditions.*Microbiological Research*. 2008;163(3):323–328.

Andrews, H. L. and Baumler, A. J. 2005. *Salmonella* species. In Fratamico, P. M., Bhunia, A. K. and Smith, J. L. (Eds.). Foodborne pathogens: Microbiology and molecular biology, p. 327-339. United Kingdom: Horizon Scientific Press Ltd.

Ben Kahla-Nakbi Amel, Besbes Amine, Chaieb Kamel, Rouabhia Mahmoud, Bakhrouf Amina. 2007. Survival of *Vibrio alginolyticus* in Seawater and Retention of Virulence of Its Starved Cells. *Marine Environmental Research* 64, 4 .

Bakhrouf A, Jeddi M, Boudabbous A, Gauthier MJ (1989) Evolution of *Pseudomonas aeruginosa* cells toward a filterable stage in seawater. *FEMS Microbiol Lett* 59:187–190.

Colwell, R.R., P.R. Brayton, D. Herrington, B.D. Tall, A. Huq, and M.M. Levine. 1996. Viable but nonculturable *Vibrio cholera* O1 revert to a culturable state in human intestine. *World J. Microb. Biotechnol.* 12, 28-31.

Costerton, J.W.; Stewart, P.S.; Greenberg, E.P. (1999). Bacterial biofilms: a common cause of persistent infections. *Science*. 284 (5418), 1318-1322.

Cappelier, J.M., Minet, J., Magras, C., Colwell, R.R. and Federighi, M. 1999. Recovery in embryonated eggs of viable but nonculturable *Campylobacter jejuni* cells and maintenance of ability to adhere to HeLa cells after resuscitation. *Appl. Environ. Microbiol.* 65:5154- 5157.

Characklis, W.G. (1990). Biofilm processes. *In* W. G. Characklis and K. C. Marshall (ed.), Biofilms. John Wiley & Sons, New York, N.Y. p. 195-231.

Dhiaf Amel, Jihen chriaa and Bakhrouf Amina. Adaptation of *Salmonella* to extreme seawater stress. Proceeding of the fifth international conference of the Mediterranean coastal environment. page.760-766. 2001.

Dhiaf Amel and Bakhrouf Amina. Fate of *Salmonella* typhimurium on rosemary and barley grown in fields treated with contaminated irrigation water *African Journal of Microbiology Research* Vol. 1(4) pp. 042-045. 2007.

Dhiaf Amel and Amina. Bakhrouf. Recovery in embryonated chicken eggs of viable but non-culturable *Salmonella*. *Journal of Food, Agriculture and Environment* Vol.2 (2) : 104-107. 2004b.

Dhiaf Amel, Neziha Chnity and Amina Bakhrouf. Fate of *Salmonella* in baby powder milk. *Journal of Food, Agriculture and Environment* Vol.2 (2) : 149-150. 2004a.

Dhiaf Amel, Bakhrouf Amina. Resuscitation of Seventeen-year Stressed *Salmonella typhimurium*. Oceanological and Hydrobiological Studies. Vol. XXXVII, No.1 (1-14) 2008.

Hamouda Elabed, Amina Bakhrouf, Rim Hamza, Maha Azaiez, Kamel Gaddour 2012 Evidence of the adaptive response in *Pseudomonas aeruginosa* to 14 years of incubation in seawater. *Journals of microbiology* .62 December 2012, Volume 62, pp 1385-1394.

Huisman GW, Siegele DA, Zambrano MM, Kolter R (1996) Morphological and physiological changes during stationary phase. In: Neidhardt FC, Curtiss R III, Ingraham JL, Lin ECC, Low KB, Magasanik B, Reznikoff WS, Riley M, Schaechter M, Umbarger HE (eds) *Escherichia coli* and *Salmonella*: cellular and molecular biology, 2nd edn. ASM Press, Washington D.C., pp 1672–1682.

Hussong D, Colwell RR, Obrien M, Weiss E, Pearson AD, Weiner RM & Burge WD (1987) Viable *Legionella pneumophila* not detectable by culture on agar media. *Bio/Technol.* 5: 947–950.

Jiang X, Chai TJ (1996) Survival of *Vibrio parahaemolyticus* at low temperatures under starvation conditions and subsequent resuscitation of viable nonculturable cells. *Appl. Environ. Microbiol.* 62:1300–1305.

Kalchayanand N., Dunne P., Sikes A., Ray B. (2004). Viability loss and morphological change of foodborne pathogens following exposure to hydrostatic pressures in the presence and absence of bacteriocins. *Int. J. Food Microbiol.*, 91: 91-98.

Lindbäck, T. M. E. Rottenberg, S. M. Roche, and L. M. Rørvik, thee ability to enter into an avirulent viable but non-culturable (VBNC) form is widespread among Listeria monocytogenes isolates from salmon, patients and environment," *Veterinary Research*, vol. 41, no. 1, 2010.

Marshall, K.C.; Stout, R.; Mitchell, R. (1971). Mechanism of the initial events in the sorption of marine bacteria to surfaces. *J. Gen. Microbiol.* 68, 337-348.

Morita, R. Y. 1993. Bioavailability of energy and the starvation state, p. 8-16. *In* S. Kjelleberg (ed.), Starvation in bacteria. Plenum Press, Inc., New York, N.Y.

Mizunoe, Y., S.N. Wai, T. Ishikawa, A. Takade, and S.-i. Yoshida. 2000. Resuscitation of viable but nonculturable cells of *Vibrio parahaemolyticus* induced at low temperature under starvation. *FEMS Microbiol. Lett.* 186,115-120.

Oliver JD, Hite F, McDougald D, Andon NL & Simpson LM(1995) Entry into, and resuscitation from, the viable but nonculturable state by *Vibrio vulnificus* in an estuarine environment. *Appl. Environ. Microbiol.* 61: 2624–2630.

Oliver, J.D. and Bockian, R. 1995. In vivo resuscitation, and virulence towards mice, of viable but nonculturable cells of *Vibrio vulnificus*. *Appl. Environ. Microbiol.* 61:2620-2623.

Oliver, J. D. "Recent findings on the viable but nonculturable state in pathogenic bacteria," *FEMS Microbiology Reviews*, vol. 34, no. 4, pp. 415–425, 2010.

Oliver, J.D. 1993. Formation of viable but nonculturable cells, p. 239-272. *In* S. Kjelleberg (ed.), Starvation in Bacteria. Plenum Press, NY.

Ophir, T.; Gutnick, D.L. (1994). A role for exopolysaccharides in the protection of microorganisms from desiccation. *Appl. Environ. Microbiol.* 60 (2), 740-745.

Pommepuy, M., Butin, M., Derrien, A., Gourmelon, M., Colwell, R.R. and Cormier, M. 1996. Retention of enteropathogenicity by viable but nonculturable *Escherichia coli* exposed to seawater and sunlight. *Appl. Environ. Microbiol.* 62:4621-4626.

Roszak D.B., Colwell R.R. (1987). Survival strategies of bacteria in the natural environment. Microbiol. Rev., 51: 365-379.

Steinert, M, Emödy, L, Amann, R, and Hacker, J, 1997, "Resuscitation of viable but nonculturable *Legionella pneumophila* Philadelphia JR32 by *Acanthamoeba castellanii*". *Appl. Environ. Microbiol.* 63, pp. 2047-2053.

Van Houdt, R.; Aertsen, A.; Jansen, A.; Quintana, A.L.; Michiels, C.W. (2004). Biofilm formation and cell-to-cell signalling in Gram-negative bacteria isolated from a food processing environment. *J. Appl. Microbiol.* 96 (1), 177-184.

Votyakova, TV, Kaprelyants, AS, and Kell, DB, 1995, "Influence of viable cells on the resuscitation of dormant cells in *Micrococcus luteus* cultures held in an extended stationary phase: the population effect". *Appl. Environ. Microbiol.* 1994. 60, pp. 3284–3291.

Villarino A., Bouvet O.M.M., Regnault B., Martin-Delautre S., Grimont P.A.D. (2000). Exploring the frontier between life and death in *Escherichia coli*: evaluation of different viability markers in live and heat or UV-killed cells. *Res. Microbiol.*, 151: 755-768.

INDEX

B

C

D

E

H

I

J

K

L

M

N

O

P

Q

R

S

T

U

V

W

Y

Z